BASIC ALGEBRAIC SYSTEMS

An Introduction to Abstract Algebra

BASIC ALGEBRAIC SYSTEMS

An Introduction to Abstract Algebra

RICHARD LAATSCH
Associate Professor of Mathematics
Miami University
Oxford, Ohio

McGRAW-HILL BOOK COMPANY

New York St. Louis
San Francisco Toronto
London Sydney

COVER PHOTOGRAPH: Sobrino, *Unstable Transformation: Juxtaposition, Superposition D*, plexiglas sculpture, courtesy Galerie Denise René, Paris.

BASIC ALGEBRAIC SYSTEMS

An Introduction to Abstract Algebra

Library of Congress Catalog Card Number 68-12266

35721

234567890 MAMM 7543210698

To ***Professors Brown*** *and* ***Hemphill***
who, in the midst of responsibilities,
took time for their students

PREFACE

In writing this introductory text in abstract algebra, I have attempted to achieve several goals; the use of earlier versions of the manuscript in the classroom has indicated some success with each of those goals. An attempt has been made to present a self-contained development of the introductory theory of groups, rings, and fields aimed at the usual variety of students found in an undergraduate mathematics class—including prospective secondary-school teachers. I have tried to avoid the time-consuming task of constructing the number systems of elementary algebra in order that more time might be devoted to the abstract theory. However, the postulates for the number systems are presented, and these systems are referred to frequently as examples. Also, the development of the theory has been tailored to help the reader analyze the familiar properties of the number systems.

In order that the material be adaptable to classes with varying mathematical backgrounds, the relevant foundational material is considered first, and the development becomes slowly more theoretical as the early chapters unfold. There are no material prerequisites for reading this book beyond advanced high school algebra, but the maturity acquired by a student in a semirigorous calculus course is assumed.

An attempt has been made to correlate those parts of algebra considered in detail here with the study of linear algebra and vector spaces. This is the specific goal of Chapter 7. This book, designed for a one-semester course, can be used together with one of the several good introductory texts in linear algebra to provide a year-long course in algebra for the undergraduate student.

My own feeling about this book is that the first five chapters constitute the heart of a first course in nonlinear abstract algebra, and the final three chapters exist to supplement, enrich, and extend the first five.

In Chapter 1 the necessary ideas of sets, relations, mappings, and operations are presented. This chapter should be read, however quickly, by even the best prepared student, since part of its purpose is to establish a uniform notation and style for the later chapters. Students requiring a fuller knowledge of the concept of a mathematical proof may skip to Chapter 8, especially Section 8-B, after the idea of an operation table is introduced in Section 1-E.

Chapters 2 and 3 on group theory have as their goal the proof of the fundamental theorem of group homomorphisms. The march to this goal is slow, with many examples and preliminary theorems along the way. Chapter 6, which develops three topics in group theory, may be studed immediately after Chapter 3, but the exposition in Chapter 6 is somewhat more sophisticated than that of any of its predecessors.

The results of Chapters 2 and 3 are used in Chapter 4 to develop properties of rings, integral domains, and fields (and the integers and rational numbers). This development continues in Chapter 5, where polynomials are considered. Here only brief mention is made of the more elegant, but less natural, representation of a polynomial as an infinite sequence. The introduction of the concept of a polynomial function helps to keep the discussion mathematically rigorous. The goals of this chapter are the unique factorization theorem for polynomials over a field and a characterization of the complex numbers.

The purpose of Chapter 7 is to relate the earlier material to the study of vector spaces. Despite a few references to Chapters 5 and 6, this chapter actually depends on Chapter 4. A short discussion of operators on groups is followed by a definition and examples of vector spaces and a discussion of the relation of some theorems about groups to vector spaces. (Such concepts as matrix algebra, determinants, dimension, and linear dependence are not developed here; my feeling is that a unified and meaningful development is possible only within the context of a course in linear algebra.) The last section of the chapter provides a point of contact with modern geometry and analysis through a discussion of nontopological topics in convexity and the geometry of linear functionals.

Three appendixes present, in order, tables for five finite groups referred to frequently as examples, a proof of the general associative law, and hints and solutions to several of the exercises in the text. (Those exercises discussed in Appendix III are indicated in the text by a star ☆.) Appendix III also contains a statement explaining the relationship of the definition of an operation to the permissibility of certain manipulations of equations.

Every theorem, definition, and exercise has two numbers. The first designates the chapter in which it occurs, and the second is its sequential number within that chapter. Thus "Theorem 3-6" denotes the sixth theorem in Chapter 3. The abbreviation "iff" is used for "if and only if," and "||" denotes the end of a proof.

An attempt has been made to be thorough in presenting those proofs that are given. However, not all theorems are accompanied by proofs; most of those proofs that are not included are requested as exercises. Alternatively, some results stated only as exercises are useful in developing subsequent material. Such exercises are indicated by being numbered in boldface type. (See, for example, Exercise 1-15.)

Mathematics is something one does, not something one just reads about, and the 460 exercises (many with several parts) are an important part of the text, providing the reader with both an opportunity to make use of the concepts of the text and an opportunity to participate in the development of the theory. A few are questions which may, in the answering, result in further questions and a view of new horizons. If an exercise is a declarative statement, a request for a proof of that statement is implied. No attempt has been made to indicate the more difficult exercises; hopefully, the reader will experience the satisfaction of discovering for himself the level of difficulty of a problem. Occasionally several exercises on a single subject appear as a group. I have used some of these groupings as an opportunity for collective class participation in searching for solutions.

As is true for most authors, I have had some much-appreciated help of several kinds from several sources. In particular, my wife has helped by being patient and understanding while I completed this book; several conversations with Charles Holmes have influenced the form of some of the presentations; and Mrs. Jean Baker has been extremely patient and conscientious while typing this and the earlier versions of the manuscript with her usual high degree of competence. I wish also to express my ultimate gratitude to the good Lord who continues to make my life both joyous and meaningful.

Richard Laatsch

CONTENTS

BASIC ALGEBRAIC SYSTEMS

An Introduction to Abstract Algebra

1 FUNDAMENTAL CONCEPTS

This chapter is devoted to those concepts of set theory needed to explain algebraic systems. The principal ideas discussed here are sets, mappings (or functions), and operations. The notation introduced here will be used consistently throughout the book.

1-A SETS

The idea of a *set* is fundamental in mathematics—so fundamental, in fact, that no attempt will be made to define it in terms of other basic concepts. However, it will be helpful if everyone is talking about approximately the same thing when the word "set" appears, so it is desirable to describe the idea in some detail. A set is a collection (flock, bunch, class, aggregate) of things such that, whenever any given thing (such as an orange or the number four) is mentioned, it is possible to decide whether that thing does or does not belong to the collection. Thus one might speak of the set of books on a particular shelf at a given time, of the set of all buildings in a city block, or of the set consisting of the numbers 2, 3, and $\sqrt{7}$. A set is thus determined by the things it does and does not contain. These "things" are called *elements* or *members* of the set, and if a "thing" x is an element of a set A, this fact is indicated by the notation

$$x \in A$$

(read: "x is an element of A"). If y is not an element of A, this is indicated by

$$y \notin A$$

This understanding of the idea of a set yields two ways of describing given sets. The simplest way is to list all of a set's elements. Braces will be used to enclose the lists. For example,

$$\{2,3,\sqrt{7}\} \qquad \{a,b,c,d\} \qquad \{\text{Thomas, Richard, Harold}\}$$

are sets. It is true that $b \in \{a,b,c,d\}$, and it is also true that $7 \notin \{a,b,c,d\}$ in the sense of the notations $\in$ and $\notin$ described above; that is, b is an element and 7 is not an element of the set $\{a,b,c,d\}$.

Many sets are too large for convenient complete listing of the elements. In such cases the "three dot" notation can be used to indicate "and so on." This device is particularly useful when the meaning of the dots can be extended to say, "and so on according to the established pattern." For example,

$$\{1,2,3,4, \ . \ . \ . \ ,499,500\}$$

would normally be interpreted as the set of all positive integers from 1 to 500 inclusive. If a last element (500) had not been indicated in the list and a set had been given by

$$\{1,2,3,4, \ . \ . \ .\}$$

then this should be interpreted as the set of all the positive integers. Thus $\{1,\frac{1}{2},\frac{1}{3},\frac{1}{4}, \ . \ . \ .\}$ would denote the set of all reciprocals of positive integers and would contain among its members such numbers as $\frac{1}{12}$, 10^{-7}, and $\frac{1}{412{,}693}$ but not the number $\frac{3}{7}$, which is not the reciprocal of any positive integer. These last two examples of sets are examples of *infinite* sets. More will be said later about the concept of infinite sets.

The description of a set obtained by listing its elements does satisfy the understanding of the idea of set because, if a set A is determined by a list of its elements, then $x \in A$ if x is on the list (or indicated to be on the list by the pattern of the list and usage of the three dots), and $x \notin A$ if x is not listed.

The second method of describing a set is that of giving a rule or test which a thing must satisfy in order to be an element of the set. For example, let set A be the set of all former Presidents of the United States as of January 1, 1965. This is written as †

$$A = \{x \mid x \text{ is a former President of the U.S. as of Jan. 1, 1965}\}$$

In general the notation is

$$\{x \mid P(x)\}$$

where P is a statement (sometimes called an *open sentence*) which must be true about x in order for x to be an element of the set. The symbol $\{x \mid P(x)\}$ is read as: "The set of *all* x such that $P(x)$ is true." This notation is frequently called the *set builder* notation. It is particularly helpful in defining infinite sets. Thus the set $\{1,\frac{1}{2},\frac{1}{3},\frac{1}{4}, \ldots\}$ already mentioned can be denoted

$$\left\{x \mid x = \frac{1}{n} \text{ and } n \text{ is a positive integer}\right\}$$

or, less formally,

$$\left\{\frac{1}{n} \mid n \text{ is a positive integer}\right\}$$

A basic relationship between sets is that of *inclusion*. By inspection, the sets $A = \{1,2,3\}$ and $B = \{0,1,4,3,7,2\}$ are revealed to have all elements of A in common. That is, every element of A is an element of B. Whenever this happens, A is called a *subset* of B.

† The sign "=" deserves a cautionary note. We use it for situations involving things other than numbers. In general, whatever p and q represent, "$p = q$" means that p and q are symbols for exactly the same thing.

Definition 1-1

A set A is a **subset** of set B iff† every element of A is also an element of B. If A is a subset of B, write

$$A \subset B$$

and write $A \not\subset B$ otherwise. (Formally, $A \subset B$ iff, for all x, $x \in A$ implies $x \in B$.)

If G and H are two sets, to prove that $G \subset H$ using the definition, it is necessary to show that *every* element of G belongs to H (equivalently, every element not in H is also not in G); but to show $G \not\subset H$, one needs only to exhibit a single element of G which is not in H. Thus $\{0,1\} \not\subset \{1,2,3\}$ because $0 \in \{0,1\}$ but $0 \notin \{1,2,3\}$. Note that it is always true that $A \subset A$ (A is a subset of itself)—as a check of the definition verifies. If $A \subset B$ and $A \neq B$, then A is called a *proper* subset of B.

In practice one always wants the sets he is discussing to be subsets of a particular set, which is called the *universal set* for that discussion.

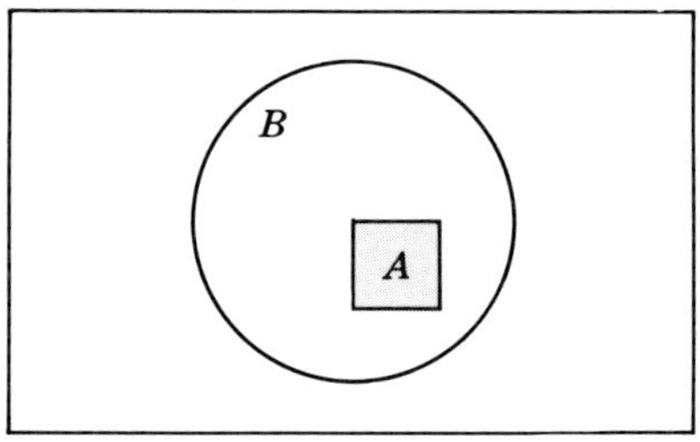

Figure 1-1 Venn diagram of $A \subset B$.

The advanced theory of sets actually requires significant restrictions on the nature of the universal set, but we regard it here chiefly as a convenience, so that, when we are talking about sets of points in a plane, for example, it will not be necessary to apply the "set builder" rules to snowflakes, cars, people, etc.

It is possible to draw useful pictures to aid the understanding of concepts about sets by letting the universal set be the set of points on and inside a rectangle and letting the sets under discussion be sets of points on and inside boundaries of curves (circles, etc.) inside the rectangle. According to this scheme the subset relation can be pictured as in Figure 1-1. Such pictures are called *Venn diagrams.*

† As stated in the preface, "iff" is used for "if and only if."

Before leaving the basic idea of a set for a consideration of combinations of sets, let us say a few words about very small sets—which can often be the most confusing kind. First, a set may have only one element. There may be only one book on a given shelf, for instance. The set $\{x \mid x = 5\}$ also has only one element. It is then easy, of course, to make a list. This is done as before by writing "$\{5\}$" or "$\{a\}$." But note that $a \in \{a\}$, whereas $a \neq \{a\}$ and $a \not\subset \{a\}$. That is, a is an element of the universal set, but $\{a\}$ is a subset of it. (You would have less trouble bouncing a ball than bouncing a set whose only element is a ball.) Also note that $a \in \{a,b\}$, but $\{a\} \subset \{a,b\}$.

In another case, if one considers the set of all students in the course Physics 472 but nobody enrolls in the course, then the class list the registrar's computing machine puts out would look, in list notation, like this:

$$\{\quad\}$$

This still fits the concept of a set because one can still say that x belongs to the set if, and only if, x is on the list. Thus there is a set with no elements, the *empty set*, which is denoted by the symbol $\varnothing$. This set also arises when using the set builder notation. For example,

$$\{\ \} = \varnothing = \{x \mid x \neq x\} = \{x \mid x > 0 \text{ and } x^2 = 0\}$$

Exercises 1-1 to 1-4

1-1. Consider the set $A = \{x \mid x \text{ is an even integer greater than } 4\}$. Which of the following are elements of A: 12, -10, $\sqrt{16}$, 5^2, $\sqrt{11}$, $\sqrt[3]{5/2}$, 2^9, 10^{-2}, π, $7^4 + 1$?

1-2. (For students of American history) Consider the set $B = \{x \mid x$ signed the Declaration of Independence$\}$. Which of the following are elements of B: George Washington, John Hancock, George III, Thomas Jefferson, Pheidippides, Button Gwinnett?

1-3. Consider the following sets:

$$A = \{1,2,3,4,5\} \qquad B = \{0,1,2,4\} \qquad C = \{1,3,5\}$$
$$D = \varnothing \qquad E = \{2\} \qquad F = \{0,2,4\}$$

List all true statements of the form $X \subset Y$ for these six sets.

1-4. Justify the following: The empty set $\varnothing$ is a subset of every set.

1-B OPERATIONS ON SETS

Numbers are useful partly because they can be combined in various ways to yield other numbers (addition, multiplication, etc.). Similarly, much of the usefulness of sets results from ways in which they can be combined to form other sets. Let us consider four operations on sets. (Just what an operation is, from a technical point of view, will be discussed soon.)

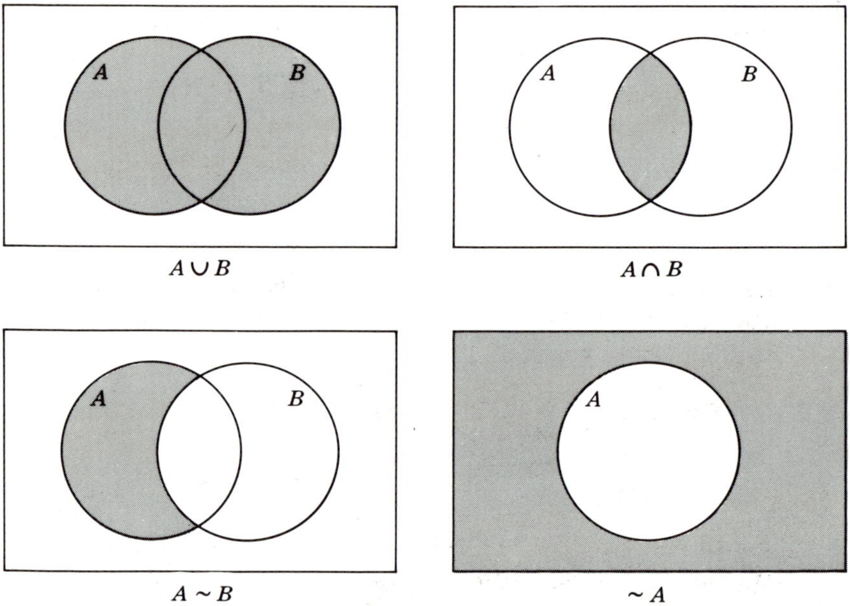

Figure 1-2 Set operations.

Definition 1-2 (See Fig. 1-2)

The **union** of two sets A and B is the set $A \cup B$ defined by

$$A \cup B = \{x \mid x \in A \text{ or } x \in B \text{ (or both)}\}$$

The **intersection** of A and B is the set $A \cap B$ defined by

$$A \cap B = \{x \mid x \in A \text{ and } x \in B\}$$

The **difference** of A and B is the set $A \sim B$ defined by

$$A \sim B = \{x \mid x \in A \text{ and } x \notin B\}$$

The **complement** of a set A is the set $\sim A$ defined by

$$\sim A = \{x \mid x \notin A\}$$

These definitions are not independent. For example, $A \sim B$ might be defined by the equation $A \sim B = A \cap (\sim B)$, and $A \cap B$ is equal to $\sim[(\sim A) \cup (\sim B)]$. Certain properties of, and relationships between, these operations are perhaps apparent. For instance

$$A \cup B = B \cup A \qquad A \cap B = B \cap A$$

$$A \cup A = A \qquad A \cap A = A$$

$$A \subset A \cup B \qquad A \cap B \subset A$$

$$A \cup (B \cup C) = (A \cup B) \cup C \qquad A \cap (B \cap C) = (A \cap B) \cap C$$

$$\sim(\sim A) = A \qquad A \sim B = A \sim (A \cap B) \qquad A \sim A = \varnothing$$

If we let $A = \{1,3,5,7,9,11\}$ and $B = \{1,2,4,5,7,9\}$ in the universal set $V = \{1,2,3, \ldots ,12\}$, then

$$A \cup B = \{1,2,3,4,5,7,9,11\} \qquad \sim A = \{2,4,6,8,10,12\}$$

$$A \cap B = \{1,5,7,9\} \qquad A \sim B = \{3,11\} \qquad B \sim A = \{2,4\}$$

If $C = \{4,6,8\}$, then $A \cap C = \varnothing$, the empty set. It is true of this example that

$$A \cup B = [(A \sim B) \cup (B \sim A)] \cup (A \cap B)$$

Is this true of any two sets A and B? (Perhaps Venn diagrams will reveal the answer.)

There will be an occasional need to consider the union or intersection of many sets at once. For this purpose an extension of parts of Definition 1-2 is required.

Definition 1-3

Let I be a set. Let A_t be a set for each $t \in I$. Then the **union** of the collection of sets $\{A_t \mid t \in I\}$ is the set

$$\bigcup_{t \in I} A_t = \{x \mid x \in A_t \text{ for at least one } t \in I\}$$

and the **intersection** of the collection is the set

$$\bigcap_{t \in I} A_t = \{x \mid x \in A_t \text{ for every } t \in I\}$$

The set I is called the *index set.*

These definitions clearly agree with Definition 1-2 in case I is a set of two elements. The index set I may be any set. (The case $I = \varnothing$ may be of some interest, although of no importance in what follows.)

A collection of sets $\{A_t \mid t \in I\}$ is called *disjoint* iff $A_i \cap A_j = \varnothing$ whenever $i \neq j$ in I. That is, a collection of sets is disjoint iff no two sets in the collection have an element in common.

Three particular sets—sets of numbers—will be used so many times in what follows that it will be worthwhile at this point to assign special symbols to be used to denote them alone.

The symbol	*Denotes the set of all*
J	Integers
Q	Rational numbers
R	Real numbers

In addition, a subscript + denotes the subset of positive elements and a subscript 0 the subset of nonnegative elements of the corresponding set J, Q, or R. For example, $R_+ = \{x \in R \mid x > 0\}$ and $J_0 = \{0,1,2,3, \ldots\}$. Note that $J_0 = J_+ \cup \{0\}$ (and similarly for Q and R).

Exercises 1-5 to 1-15

1-5. Describe (in words) the following sets:

$$R_+ \sim Q \qquad Q \sim R \qquad J \sim J_0 \qquad Q \cap R_+ \qquad Q \sim J$$

1-6. Using the sets of Exercise 1-3, find:

a. $A \cup B$ b. $A \cap B$ c. $A \cup C$
d. $B \cap B$ e. $A \sim B$ f. $D \cap E$
g. $F \sim A$ h. $J_+ \sim B$ i. $B \sim J_+$

1-7. Draw Venn diagrams for $(\sim A) \cup (\sim B)$ and $\sim(A \cap B)$. Are these the same set?

1-8. Use Venn diagrams to determine which pairs of the following sets are equal:

$$A \sim (B \sim C) \qquad (A \sim B) \sim C \qquad A \sim (B \cup C)$$
$$A \sim (B \cap C) \qquad (A \sim B) \cup (A \cap C)$$

1-9. Use Venn diagrams to verify the following equations:
a. $(A \sim B) \cup (A \cap B) = A$
b. $(A \cup B) \sim A = B \sim A$
c. $A \cup (B \cap C) = (A \cup B) \cap (A \cup C)$
d. $A \cap (B \sim A) = \varnothing$

1-10. List the elements of the following sets:

$$A = \{x \mid x \in J_+ \text{ and } x < 8\}$$
$$B = \left\{x \mid x = \frac{1}{n} \text{ and } n \in J_+ \text{ and } 3 < n < 5\right\}$$
$$C = \{x \mid x \text{ is an even negative integer}\}$$

1-11. Can a collection of sets be disjoint and not have an empty intersection? Can it have an empty intersection and not be disjoint?

1-12. (Russell paradox) Call a set A *normal* iff A is not an element of itself. (Example: The set of all apples is not an apple.) Call A *abnormal* otherwise. (The set of all concepts is a concept.) Consider the set N of all normal sets. Is N a normal or an abnormal set?

1-13. Use the definitions to prove
a. $A \cap B \subset B$
b. $A \cup B = B$ iff $A \subset B$
☆c.† $A \cap (B \cup C) = (A \cap B) \cup (A \cap C)$
d. $A \cup (B \cap C) = (A \cup B) \cap (A \cup C)$
e. $A \sim B = A \cap (\sim B)$
(Note with respect to *c*, *d*, and *e* that $X = Y$ iff $X \subset Y$ and $Y \subset X$. Thus "=" can be proved by proving that each set is a subset of the other.)

† Here and elsewhere the "☆" indicates that a hint or solution to the exercise is given in the back of the book.

1-14. Prove that

a. $\sim(\bigcup_{t \in I} A_t) = \bigcap_{t \in I} (\sim A_t)$

b. $\sim(\bigcap_{t \in I} A_t) = \bigcup_{t \in I} (\sim A_t)$

c. $A \cap (\bigcup_{t \in I} B_t) = \bigcup_{t \in I} (A \cap B_t)$ (Compare with Exercise 1-13c.)

1-15. a. Show by Venn diagrams or by manipulations based on previous results that

$$(A \sim B) \cup (B \sim A) = (A \cup B) \sim (A \cap B)$$

b. Define an operation Δ on sets by

$$A \,\Delta\, B = (A \sim B) \cup (B \sim A)$$

The operation Δ is called the *symmetric difference* of A and B. Use Venn diagrams to verify that $A \,\Delta\, B = B \,\Delta\, A$ and $A \,\Delta\, (B \,\Delta\, C) = (A \,\Delta\, B) \,\Delta\, C$.

c. Verify by definitions that $A \,\Delta\, \varnothing = A$ and $A \,\Delta\, A = \varnothing$.

1-C RELATIONS

Other basic concepts of mathematics are those of *relations* and *mappings*. These depend upon the idea of an *ordered pair*. An ordered pair of elements x and y will be denoted by "(x,y)" with the understanding that $(x,y) = (y,x)$ if, and only if, $x = y$. Thus it is of some concern whether x or y is listed first. [A formal definition in terms of sets can be given by the equation "$(x,y) = \{\{x\},\{x,y\}\}$," which serves to distinguish x from y, but the idea of an ordered pair is sufficiently commonplace to serve as a starting point for the definition of relation.]

Definition 1-4

Let A and B be sets. The **cartesian product** of A and B is the set $A \times B$ defined by

$$A \times B = \{(a,b) \mid a \in A \text{ and } b \in B\}$$

Notice that, since the definition deals with ordered pairs, $A \times B \neq B \times A$ in general. If A and B are the subsets of the set R of real numbers defined by

$$A = \{x \mid 1 \leq x \leq 4\} \qquad \text{and} \qquad B = \{x \mid 2 \leq x \leq 3\}$$

$A \times B$ can be pictured (as a subset of $R \times R$) as in Figure 1-3.

The actual definition of *relation* is an almost total abstraction. It will be desirable to discuss the idea informally first in order to show the definition actually does give what is wanted.

The word "relation" is most readily applied to people for such specific relations as "Jim is the brother of Sam," "Smith is the employer of McGrath," and "Brown is the neighbor of Henderson." A more

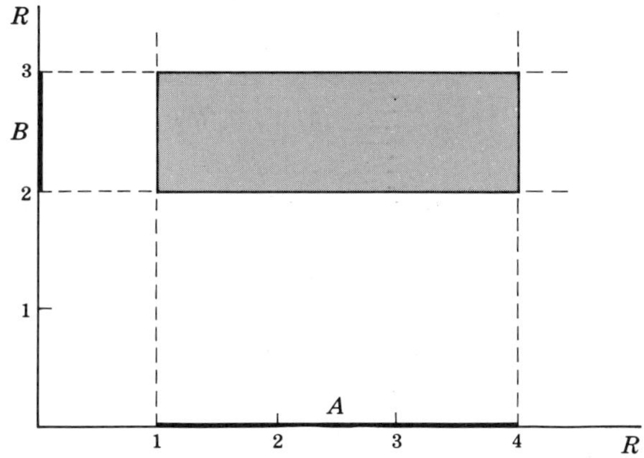

Figure 1-3 $A \times B$.

mathematically oriented example of a relation is one such as "five is less than eleven." Thus a relation is specified in practice by words such as "employer of" or "less than." Once such a description exists, the problem is to determine what pairs of things are thus related. Alternatively, one might begin by knowing what pairs are related and then seek a "name" for the relation. It is this second attitude which dominates in the definition, although the first point of view is appropriate for many practical situations.

Definition 1-5

A **relation** on A to B is a subset of $A \times B$. [That is, a relation on A to B is some set of ordered pairs (a,b) with $a \in A$ and $b \in B$.] If

ρ is a relation on A to B, write

$$a \rho b \text{ iff } (a,b) \in \rho$$

and $a \not\rho b$ otherwise. The **domain** of ρ is the subset of A consisting of first coordinates of elements (ordered pairs) of ρ. Thus the domain is the set

$$\text{dm}(\rho) = \{a \mid a \in A \text{ and, for some } b \in B,\ (a,b) \in \rho\}$$

Similarly, the **range** of ρ is the set

$$\text{rn}(\rho) = \{b \mid b \in B \text{ and, for some } a \in A,\ (a,b) \in \rho\}$$

of all second coordinates of elements of ρ.

Definition 1-6

A relation on A to A is called a **relation in** A. (Thus a relation ρ is a relation in A iff both the domain and range of ρ are subsets of A.) Three important potential properties of a relation ρ in a set A are:

r. For all $x \in A$, $x \rho x$ (*reflexive property*).
s. For all $x,y \in A$, if $x \rho y$, then $y \rho x$ (*symmetric property*).
t. For all $x,y,z \in A$, if $x \rho y$ and $y \rho z$, then $x \rho z$ (*transitive property*).

As examples, the relation "is a father of" has none of these properties, the relation "is less than" defined in the set J of integers is transitive but neither reflexive nor symmetric, and the relation $\{(1,1),(2,2),(3,3),(1,2),(2,1)\}$ is reflexive, symmetric, and transitive in the set $\{1,2,3\}$ (but not reflexive in $\{1,2,3,4\}$).

Definition 1-7

A relation ρ in a set A is an **equivalence relation** in A iff ρ is reflexive, symmetric, and transitive in A.

Some examples of equivalence relations are equality of real number symbols and parallelism of lines (assuming a line to be parallel to itself). The smallest equivalence relation in a set A (a subset of every other equivalence relation in A) is equality of elements; no relation could be smaller and be reflexive. The largest equivalence relation in A is $A \times A$; no relation in A whatever could be larger.

There is a difference in form between the reflexive property, on the one hand, and the symmetric and transitive properties on the other. The latter two say that if something is true, then something else is true; but the reflexive property flatly says something is true—and for all elements of A.

Exercises 1-16 to 1-25

1-16. Let $A = \{x \mid x \in R \text{ and } 0 \leq x \leq 4\}$; let $B = \{0,1,2,3\}$. Draw the graph of $A \times B$. Draw the graph of the relation $\leq$ on A to B. Describe dm ($\leq$) and rn ($\leq$).

1-17. Consider the relation $\rho = \{(a,b),(a,c),(a,a),(b,d),(d,d)\}$ in the set $\{a,b,c,d\}$.

a. Adjoin the minimum number of ordered pairs necessary to make it reflexive.

b. Adjoin the minimum number of additional ordered pairs to make it symmetric.

c. Adjoin the minimum number of additional ordered pairs to make it an equivalence relation.

1-18. Omit (a,c) from the set ρ in Exercise 1-17 and repeat that exercise.

1-19. a. Describe the reflexive and symmetric properties for a relation in R in terms of the geometry of the coordinate plane.

b. (More difficult) Find a geometric description of the transitive property.

1-20. Each of the properties r, s, and t may be true or false independent of the other two. Give a mathematically oriented example of a set A and a relation in it for each of the eight possible combinations of truth and falsity of r, s, and t. (For example, let $A = R$; then "$x \rho y$ iff $x < y$" defines a ρ which is not r, not s, and t.)

1-21. What are the domain and range of an equivalence relation in a set A? Why?

1-22. Consider the relation $<$ in J_+. What is its domain? What is its range?

1-23. Prove that $(A \cup B) \times C = (A \times C) \cup (B \times C)$.

1-24. (A fallacy) "Let ρ be a relation in A satisfying s and t. Let $a \in A$, and let $b \in A$ such that $a \,\rho\, b$. Then $a \,\rho\, b$ implies $b \,\rho\, a$ (by s), and $a \,\rho\, b$ and $b \,\rho\, a$ imply $a \,\rho\, a$ (by t). Therefore, if ρ is symmetric and transitive, it is also reflexive." Criticize this argument. (Note that it is inconsistent with the opening sentence of Exercise 1-19.)

1-25. Let $A = \{a,b,c,d\}$. Let

$$\rho = \{(a,a),(a,b),(b,b),(b,a),(c,d),(c,c),(d,d),(b,c),(a,c),(b,d)\}$$

Is ρ reflexive, symmetric, transitive in A? Give a reason for each answer.

▲

The study of equivalence relations in a set A can be identified completely with the study of certain kinds of collections of subsets of A.

Definition 1-8

A *disjoint* collection of nonempty subsets of A whose union is A is called a **partition** of A. That is, P is a partition of A iff

1. $X \in P$ implies $X \subset A$ and $X \neq \varnothing$,
2. $X,Y \in P$ and $X \neq Y$ imply $X \cap Y = \varnothing$, and
3. $\bigcup_{X \in P} X = A$.

For example, the set of even integers and the set of odd integers form a partition of the set J of all integers. Also $\{\{0\},\{1,-1\},\{2,-2\},\{3,-3\}, \ldots\}$ is a partition of J. The collection $\{\{1,2\},\{3,4\},\{5,6\}, \ldots\}$ is a partition of J_+.

The most significant fact about partitions is that each partition determines a unique equivalence relation and each equivalence relation determines a unique partition.

Theorem 1-1

Let ρ be an equivalence relation in a set A and let the set E_a be defined for each $a \in A$ by

$$E_a = \{x \mid x \in A \text{ and } x \,\rho\, a\}$$

Then the collection P of all sets E_a is a partition of A. Conversely, if P' is any partition of A, then the relation σ defined by "$x \sigma y$ iff x and y are in the same member of P'" is an equivalence relation in A.

DISCUSSION. The first part of the theorem is not trying to say that a and b determine different sets E_a and E_b, but that they determine the same set if $a \rho b$ and disjoint sets if $a \not\rho b$. Since the proof of the second part of the theorem is trivial, only the first part will be proved here.

PROOF. To show that P is a partition, let us first note that $E_a \subset A$ for every $a \in A$ because of the way E_a is defined and that $E_a \neq \varnothing$ since $a \in E_a$ by the reflexive property of ρ. Also $\bigcup_{a \in A} E_a = A$ since there is a set E_a for every $a \in A$. To complete the proof that P is a partition of A, let us show that different sets E_a have no common element—by showing that, if E_a and E_b do have a common element, they are equal.

Assume $E_a \cap E_b \neq \varnothing$. Let $c \in E_a \cap E_b$. Then $c \rho a$ by definition of E_a and $c \rho b$ by definition of E_b. We show first that $E_a \subset E_b$. To this end, let $x \in E_a$. Then $x \rho a$; $c \rho a$ implies $a \rho c$ (by symmetry); and $x \rho a$ and $a \rho c$ imply $x \rho c$ (by transitivity). Further, $x \rho c$, and $c \rho b$ imply $x \rho b$ (by transitivity again), so $x \in E_b$ by the definition of E_b. Since every element of E_a is in E_b, $E_a \subset E_b$. By the same sort of argument $E_b \subset E_a$; thus $E_a = E_b$. ||

Definition 1-9

The sets E_a defined in Theorem 1-1 are called the **equivalence classes** determined by the equivalence relation ρ in set A.

Exercises 1-26 to 1-30

1-26. Let $A = \{0,1,2, \ldots ,25\}$. Define the relation ρ in A by: $a \rho b$ iff a and b have the same remainder (0, 1, 2, or 3) when divided by 4. Prove that ρ is an equivalence relation in A and list the equivalence classes.

1-27. How many equivalence relations can be defined in $A = \{1,2,3\}$?

1-28. How many subsets does a set of n elements have? How many relations can be defined in a set of n elements?

1-29. Describe the equivalence classes for the relation of parallelism of lines in a plane.

1-30. List all partitions of the following sets.
a. $\{1,2\}$ b. $\{1,2,3\}$ (See Exercise 1-27)
c. $\{1,2,3,4\}$

1-D MAPPINGS

The notion of a relation makes it possible to define the idea of a function, or mapping, quite easily since it is possible to think of mappings, too, in terms of ordered pairs. The intuitive notion of a mapping is that of taking elements of some set A and identifying them individually with elements of a set B. For example, the equation $y = 6x + 2$ pairs each number x with some number y. Actually, it is often convenient to think of a mapping as "doing something" to each element of A (such as transforming it into an element of B).

Definition 1-10

Let A and B be nonempty sets. A **mapping** (or **function**) α from A to B is a relation on A to B such that *every* element $a \in A$ is used *exactly once* as the first coordinate of an ordered pair in α. If α is a mapping from A to B, write $\alpha : A \to B$. If $(a,b) \in \alpha$ (that is, $a\ \alpha\ b$), write $b = a\alpha$ or $\alpha : a \to b$.

Thus α is a mapping from A to B iff (1) dm $(\alpha) = A$ and (2) $b = a\alpha$ and $c = a\alpha$ imply $b = c$; a mapping is a specialized kind of relation. The notation $a\alpha = b$ is, of course, a departure from that used in introductory mathematics courses such as calculus. The usage here is $a\alpha$ instead of $\alpha(a)$ [or xf instead of $f(x)$] to denote the second coordinate of a member of a mapping. Frequently the element $a\alpha$ in B is referred to as the *value* of α at a or the *image* of a under α. Actually there are a variety of notations available for use, such as $y = x\alpha$ (mapping notation), $x\ \alpha\ y$ (relation notation), and $(x,y) \in \alpha$ (set notation), to denote the same thing. The notation $y = x\alpha$ will be used here, largely because it gives a more convenient way of thinking about combinations of mappings. Also figurative language such as,

"Let α map A to B," should be referred to the idea of Definition 1-10 when any need arises to be careful in discussing a mapping.

The definition of mapping pays very little attention to the set B. All that is known is that B contains the range of α, the set of second coordinates of members of α. In fact there is nothing in Definition 1-10 to keep rn (α) from being a one-point subset of B. If this is the case, α is called a *constant mapping*. As a simple finite example of a mapping, let $A = \{1,2,3\}$ and $B = \{a,b,c\}$; then $\alpha = \{(1,a),(2,c),(3,a)\}$ is a mapping from A to B with dm $(\alpha) = A$ (naturally) and

$$\text{rn }(\alpha) = \{a,c\} \subset B$$

However, $\beta = \{(1,a),(2,c),(3,a),(1,b)\}$ is a relation which is not a mapping since 1 is the first coordinate of two different pairs. In the first case, that of α, one can write $1\alpha = a$, $2\alpha = c$, and $3\alpha = a$. [The insistence upon dm $(\alpha) = A$ is somewhat artificial since, as long as no member of A is used two or more times, a nonempty subset of $A \times B$ is a mapping from the subset of A consisting of those elements which *are* first coordinates.]

Because of the amount of freedom allowed by the definition of mapping, special words are used to describe certain important kinds of mappings.

Definition 1-11

Let α be a mapping from A to B. Then α is called a mapping **onto** B iff rn $(\alpha) = B$ (i.e., iff every element of B is a second coordinate in α). Also, α is a **one-to-one** mapping iff every element of rn (α) occurs exactly once as a second coordinate in α (that is, iff $z = x\alpha$ and $z = y\alpha$ imply $x = y$).

These two terms can be combined when desirable to talk about a *one-to-one* mapping from A *onto* B—which would be a mapping α that identifies each $a \in A$ with exactly one $b \in B$ and each $b \in B$ with exactly one $a \in A$. Every mapping is a mapping onto some set, namely the range of the mapping, but one must always be careful to see whether "onto" is, or is not, used in describing a mapping.

Definition 1-11 is a constructive statement in that it determines ways of proving that mappings are "onto" or one-to-one. To show α is a mapping from A onto B, let b be an arbitrary element of B and show that there is some element $a \in A$ such that $a\alpha = b$. An intuitive idea of one-to-one is that different elements are paired with different elements, so that if $u \neq v$, then $u\alpha \neq v\alpha$. Since "$=$" is an easier concept to manipulate than "$\neq$," it is usually easier to work from the definition and prove that, if $u\alpha = v\alpha$, then $u = v$.

One-to-one onto mappings are the essence of counting. There are as many fingers on one normal hand as players on a basketball team because there is a one-to-one mapping of the set of fingers onto the set of players—or because there is a one-to-one mapping of each set onto the set $\{1,2,3,4,5\}$.

Capitalizing on the observation above that a one-to-one mapping from A onto B identifies different elements of A with different elements of B, and vice versa, one might observe that by reversing the first and second coordinates in the pairs $(a,b) \in \alpha$ he obtains a set of ordered pairs (b,a) *which is still a mapping*. This mapping is called the *inverse* of α and denoted by α^{-1}.

Definition 1-12

If α is a one-to-one mapping from A onto B, the **inverse** of α is the mapping α^{-1} defined by

$$\alpha^{-1} = \{(b,a) \mid (a,b) \in \alpha\}$$

The domain and range of α^{-1} are B and A, respectively—just the opposite of the domain and range of α. (The general concept of the inverse of a relation is mentioned in the following exercises but is not needed later in this book.)

One particular way of combining mappings is of particular importance in all areas of mathematics. That way is called *composition*.

Definition 1-13

Let $\alpha : A \to B$ and $\beta : B \to C$. The **composite** of α and β is the mapping $\alpha\beta : A \to C$ defined by

$$x(\alpha\beta) = (x\alpha)\beta \text{ for all } x \in A$$

The domain of $\alpha\beta$ is A, and the range of $\alpha\beta$ is a subset of rn (β) in C. The image in C of an element $x \in A$ is determined by first finding $x\alpha$ in B and then applying β to that element to obtain $(x\alpha)\beta = x(\alpha\beta)$

in C (Figure 1-4). Of course, nothing prevents A, B, and C from being the same set, and this special case is one of the most important for applications of composition. [It is the fact that mappings can be

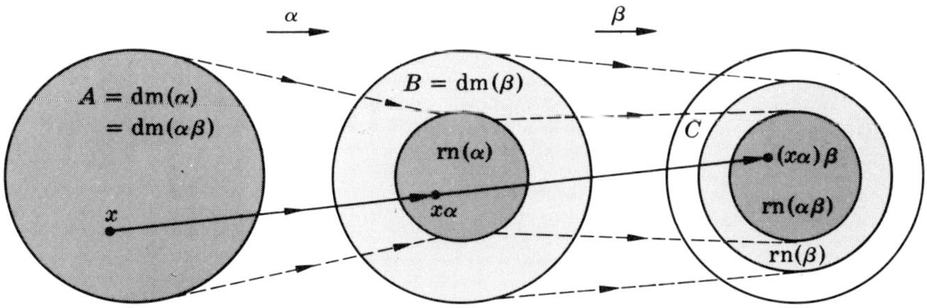

Figure 1-4 The composite $\alpha\beta$.

written in the order in which they are applied which is the principal motivation for using $x\alpha$ in preference to $\alpha(x)$ as a notation.]

Theorem 1-2

If $\alpha : A \to B$, $\beta : B \to C$, and $\gamma : C \to D$ are mappings, then the mappings $\alpha(\beta\gamma)$ and $(\alpha\beta)\gamma$ are both defined and are equal.

DISCUSSION. The mapping $(\alpha\beta)\gamma$, for example, means that $\alpha\beta$ is applied to elements of A, yielding elements of C to which γ is applied to yield elements of D. The proof is designed to show that both mappings amount to α followed by β followed by γ.

PROOF. A brief check verifies that both $(\alpha\beta)\gamma$ and $\alpha(\beta\gamma)$ are mappings with domain equal to A and range a subset of D. The mappings will be proved equal by showing that their effect on every element of A is the same. Definition 1-13 is used repeatedly.

Let $x \in A$. Then

$$x[(\alpha\beta)\gamma] = [x(\alpha\beta)]\gamma = [(x\alpha)\beta]\gamma$$

Also $x[\alpha(\beta\gamma)] = (x\alpha)(\beta\gamma) = [(x\alpha)\beta]\gamma$. Thus $x[\alpha(\beta\gamma)] = x[(\alpha\beta)\gamma]$ for all $x \in A$. $\|$

Exercises 1-31 to 1-45

1-31. a. List all mappings of $\{1,2\}$ to $\{a,b\}$, of $\{1,2,3\}$ to $\{a,b\}$, of $\{a,b\}$ to $\{1,2,3\}$.

b. How many mappings exist from a set of n elements to a set of r elements?

☆ ***1-32.*** Prove that the composite of two onto mappings is onto. (That is, if $\alpha : A \rightarrow B$ and $\beta : B \rightarrow C$ are mappings onto B and C, respectively, then $\alpha\beta$ is a mapping onto C.)

1-33. Prove that the composite of two one-to-one mappings is one-to-one.

1-34 to 1-38. (***Infinite sets***) A nonempty set A is called an *infinite* set iff there exists a *one-to-one* mapping from A *onto* a *proper* subset of A. A set A is called *finite* iff A is not infinite. (The empty set is finite.)

1-34. Prove that, if $A \subset B$ and A is infinite, then B is infinite. *Corollary:* If A is infinite and C is any set, then $A \cup C$ is infinite.

1-35. Give an example of a one-to-one mapping of a set to itself which is not onto.

1-36. Give an example of a mapping of a set onto itself which is not one-to-one.

1-37. a. Show that $\{1,2,3,4\}$ is finite.

☆b. Characterize finite sets in terms of one-to-one mappings to the set of positive integers J_+.

1-38. Two sets A and B are said to be the *same size* iff there exists a one-to-one mapping from A onto B. Show that J and J_+ are the same size.

1-39. Let α be a one-to-one mapping from A onto B and β a one-to-one mapping from B onto C. Show that $(\alpha\beta)^{-1} = \beta^{-1}\alpha^{-1}$. [Note that $(\alpha\beta)^{-1}$ exists as a mapping by Exercises 1-32 and 1-33.]

1-40. Definition 1-10 specifies *nonempty* sets A and B. Consider, however, the nature of the ordered pairs obtained if A is empty; if B is empty.

1-41. If ρ is a relation from A to B, the *inverse* of ρ is the relation ρ^{-1} from B to A defined by

$$\rho^{-1} = \{(b,a) \mid (a,b) \in \rho\}$$

Thus $a\,\rho^{-1}\,b$ iff $b\,\rho\,a$. Give an example of a function whose inverse relation is not a function. Give the domain and range of this relation and its inverse.

1-42. Make an appropriate definition of the composite of two relations, and prove a result similar to Exercise 1-39 for relations.

1-43. Let α be a one-to-one mapping from A onto B. Show that $x(\alpha\alpha^{-1}) = x$ for all $x \in A$ and $y(\alpha^{-1}\alpha) = y$ for all $y \in B$.

1-44. Let $\alpha : A \to B$, $A_1 \subset A$, and $B_1 \subset B$. Define

$$A_1\alpha = \{b \mid b = a\alpha \text{ for some } a \in A_1\} \text{ and}$$
$$B_1\alpha^{-1} = \{a \mid b = a\alpha \text{ for some } b \in B_1\}$$

a. Prove: $(A_1 \cup A_2)\alpha = A_1\alpha \cup A_2\alpha$.
b. Prove: If $A_1 \subset A_2$, then $A_1\alpha \subset A_2\alpha$.
c. Show by example that $(A_1 \cap A_2)\alpha \neq A_1\alpha \cap A_2\alpha$.
d. Show inclusion holds one way in (c).
e. Prove: $(B_1 \cap B_2)\alpha^{-1} = B_1\alpha^{-1} \cap B_2\alpha^{-1}$. Does the corresponding statement hold for unions?
f. Is $(\sim A_1)\alpha = \sim(A_1\alpha)$?
g. Is $(\sim B_1)\alpha^{-1} = \sim(B_1\alpha^{-1})$?

1-45. Let $\alpha : A \to B$ and, for each $y \in \text{rn}\,(\alpha)$, let $\bar{y} = \{x \mid x \in A \text{ and } x\alpha = y\}$. (Note that $\bar{y}$ is a subset of A.) Show that

$$P = \{\bar{y} \mid y \in \text{rn}\,(\alpha)\}$$

is a partition of A. What is the associated equivalence relation?

1-E OPERATIONS

The usual understanding of an operation such as addition or multiplication of numbers is that two numbers are combined to yield a third number (as in $2 + 6 = 8$). The idea of an operation is not limited to numbers; the basic concept of combining two elements of a set to get a third element is the basis of the following definition.

Definition 1-14

Let A be a nonempty set. An **operation** in A is a mapping σ from

$A \times A$ to A. If $(x,y)\sigma = z$, the usual notation is

$$x \, \sigma \, y = z$$

The most-used symbols for operations will be $+$, $\cdot$, $\circ$, $*$, and juxtaposition (placing elements side-by-side, as in "xy"). In general, most language used to describe operations is borrowed from multiplication of numbers, and it is thus customary to refer to $x \circ y$ as the "*$\circ$ product*" (or just the *product*) of x and y. The elements x and y are *factors*. A "$+$ product" will still be called a *sum*.

What has been called an operation above could be more explicitly called a *binary* operation to indicate that two elements are used to determine another element. In many discussions this is done; in these cases one might also wish to consider *unary* operations (mappings from A to A such as taking the negative of a number), *ternary* operations (mappings from $A \times A \times A$ to A), and the like. In this book, however, anything called an operation will always be a binary operation.

The definition of an operation in a set A as a mapping with domain $A \times A$ means that an operation is applied to *ordered* pairs. One cannot expect $x \circ y$ to be the same as $y \circ x$ unless additional information is available to justify such an assumption. Much of the time in what follows it will be the case that $x \circ y \neq y \circ x$. For example, in subtracting real numbers, $3 - 5 \neq 5 - 3$, and, in fact, $a - b \neq b - a$ unless $a = b$.

We earlier considered the operations of union ($\cup$), intersection ($\cap$), and difference ($\sim$), as well as symmetric difference (Δ, Exercise 1-15) defined on the set of all subsets of a given set A. We shall denote this set in the future by $P(A)$. That is:

$$P(A) = \{X \mid X \subset A\}$$

The next definition provides the title for this book. Actually the definition is more complicated (because of the ". . .") than the uses of it require. The examples following will indicate the sort of specific systems to be considered later. There will be no occasion to consider anything more involved than a set, two operations, and one relation.

Definition 1-15

An **algebraic system** is a collection whose elements are a nonempty set A, some operations in A (or none), and some relations in A (or none). The notation

$$\langle A, \circ, *, \ldots ; \rho, \tau, \ldots \rangle$$

will denote a collection in which the first entry is a nonempty set, the other entries before the semicolon are operations, and all entries after the semicolon are relations. The symbol $\langle A, \circ, *, \ldots ; \rho, \tau, \ldots \rangle$ actually denotes an algebraic system if, and only if, the operations are operations *in* A and the relations are relations *in* A.

For example, (1) $\langle R, +, \cdot ; < \rangle$ denotes the system consisting of the real numbers with operations addition and multiplication and relation "less than"; (2) $\langle P(B), \cup, \cap \rangle$ denotes the system consisting of all subsets of B with operations union and intersection and no relations; (3) $\langle P(B); \subset \rangle$ denotes the system of all subsets of B with no operations and the subset relation. However, $\langle \{0,1\}, +, \cdot \rangle$ is not an algebraic system since the usual $+$ of the number systems is not an operation in $\{0,1\}$; $1+1 \notin \{0,1\}$. If $\langle A, \circ, * \rangle$ is an algebraic system, then so are $\langle A, \circ \rangle$ and $\langle A, * \rangle$.

The next definition introduces the first of several properties of operations which occur frequently and therefore constitute the main subjects of the remainder of this book.

Definition 1-16

Let $\circ$ be an operation and A a set. The operation $\circ$ is **associative** in A iff

$$x \circ (y \circ z) = (x \circ y) \circ z \text{ for all } x,y,z \in A$$

The operation $\circ$ is **commutative** in A iff

$$x \circ y = y \circ x \text{ for all } x,y \in A$$

Every effort should be made to distinguish between the properties of associativity and commutativity. They are frequently confused. To "associate" means to group, and associativity refers to regrouping of elements without change of their order; to "commute" is to interchange, and commutativity refers to interchanging the order of two elements.

If an operation $\circ$ is associative, then

$$x \circ y \circ z$$

is a well-defined symbol—meaning $x \circ (y \circ z)$ or $(x \circ y) \circ z$ without fear of error or confusion. But if $\circ$ is not associative, "$x \circ y \circ z$" is ambiguous and must be avoided since it then matters whether the first or second "$\circ$" is performed first. For example, $5 - (3 - 2) \neq (5 - 3) - 2$, so that $5 - 3 - 2$ is meaningless (unless additional rules for the subtraction operation are established). However, $6 \cdot 2 \cdot 5 = 60$ whether it is interpreted as $6 \cdot (2 \cdot 5)$ or $(6 \cdot 2) \cdot 5$.

It is true that, if an operation $\circ$ is both associative and commutative, then the operation may be performed on a set of elements independent of their order. In this case, for example,

$$a \circ (b \circ c) = b \circ (a \circ c) \qquad \text{and} \qquad (a \circ b) \circ (c \circ d) = c \circ [(b \circ d) \circ a]$$

To verify the second equation,

$$\begin{aligned} (a \circ b) \circ (c \circ d) &= a \circ [b \circ (c \circ d)] && \text{Associativity} \\ &= [b \circ (c \circ d)] \circ a && \text{Commutativity} \\ &= [(c \circ d) \circ b] \circ a && \text{Commutativity} \\ &= [c \circ (d \circ b)] \circ a && \text{Associativity} \\ &= [c \circ (b \circ d)] \circ a && \text{Commutativity} \\ &= c \circ [(b \circ d) \circ a] && \text{Associativity} \end{aligned}$$

(Does there exist a solution to the above involving fewer than six steps?) No proof of the general statement above about independence of order will be given, although theorems relating to it will appear soon.

A somewhat different, and quite important, example of an operation is provided by the concept of composition of mappings. First, let us call a mapping of a set into itself a *transformation* of the set. [That is, α is a transformation iff rn $(\alpha) \subset$ dm (α).] Next, consider a fixed nonempty set A and all the transformations of A. Let $T(A)$ denote the set of all transformations of A. If α and β are in $T(A)$, then $\alpha : A \to A$ and $\beta : A \to A$, so $\alpha\beta : A \to A$. Thus composition of mappings is an operation in the set $T(A)$, and $\langle T(A), \ \rangle$† is an algebraic system. Note that the elements in this system are the mappings on A, not the members of A. By Theorem 1-2 the operation (composition of transformations) is associative. However, it is not commutative.

† The blank space in $\langle T(A), \ \rangle$ means that the operation is denoted by juxtaposition (i.e., $\alpha\beta$ is the product of α and β).

Exercises 1-46 to 1-52

1-46. Which of the following are algebraic systems?

a. $\langle R \sim Q, +\rangle$ b. $\langle \{0,1\}, \cdot\rangle$

c. $\langle P(J \sim \{0\}), \cup, \cap\rangle$ d. $\langle P(J) \sim \{\varnothing\}, \cup, \cap\rangle$

e. $\langle J_0, -\rangle$

1-47. Let A be a set. Consider $\langle P(A), \cup, \cap\rangle$. Which of the operations are associative? Which are commutative?

1-48. What is the minimum number of steps needed to transform $(a \circ b) \circ c$ into $(c \circ b) \circ a$ if $\circ$ is associative and commutative? What is the minimum number for transforming $[(a \circ b) \circ c] \circ d$ into $[(d \circ c) \circ b] \circ a$? Generalize to a product of n elements—reversing the order and associating "to the left" both initially and finally.

1-49. Show that the transformations α and β of the real numbers R, defined by $x\alpha = x + 5$ and $x\beta = 2x$, do not commute. Find two distinct transformations of R which do commute.

1-50. Define two one-to-one transformations of $\{1,2,3\}$ onto itself which do not commute. Define two distinct ones which do commute.

1-51. Which of the operations addition, subtraction, and division are associative in R? Which are commutative?

1-52. Show that $\sim$ is not associative in $P(X)$. Is there a subset relation between the two associations of $A \sim B \sim C$? (See Exercise 1-8.)

Definition 1-17

Let $\circ$ and $*$ be operations and A a nonempty set. Then $*$ is **left-distributive** over $\circ$ in A iff

$$x * (y \circ z) = (x * y) \circ (x * z) \text{ for all } x,y,z \in A$$

If $(y \circ z) * x = (y * x) \circ (z * x)$ for all $x,y,z \in A$, then we say that $*$ is **right-distributive** over $\circ$ in A. If $*$ is both left-distributive and right-distributive over $\circ$, then $*$ is said to be **distributive** over $\circ$.

Left- or right-distributivity in a set is a relation between two operations. It is most familiar in the number systems as a relation between multiplication and addition. Thus if $a,b,c \in R$, then

$$a \cdot (b + c) = a \cdot b + a \cdot c$$

(This is sometimes referred to as *removing the common factor.*) However, addition is not left-distributive over multiplication in R. For example,

$$17 = 2 + (3 \cdot 5) \neq (2 + 3) \cdot (2 + 5) = 35$$

All of the properties considered so far—associativity, commutativity, and distributivity—are *hereditary.* That is, if they are true of an operation $\circ$ when the elements are in set A, then they are also true when the elements are taken from a subset B of A. Thus if $\circ$ is commutative in A and $a,b \in B \subset A$, then $a \circ b = b \circ a$. (This statement says nothing about whether $a \circ b$ is actually in B or not. The equality may be that of elements in some set containing B.)

Exercises 1-53 to 1-55

1-53. Is division ($\div$) left-distributive over addition in R? Right-distributive? Is multiplication left- or right-distributive over subtraction?

1-54. Consider the operations $\cup$, $\cap$, and $\sim$ in $P(A)$. Which of these is left-distributive over another? (Six pairs to consider.)

1-55. Prove: If $*$ is commutative and left-distributive over $\circ$, then $*$ is right-distributive over $\circ$.

▲

In the number systems certain numbers have special properties. The number $0 \in R$, for instance, has the property that $a + 0 = a$ for all $a \in R$. Also $a \cdot 0 = 0$ for all $a \in R$. If $b \in R$ is given and $b \neq 0$, then there is a number ${}^{-}b$ (or $-b$) such that $b + {}^{-}b = 0$. It is properties such as these which are considered next.

Definition 1-18

Let $\langle A,\circ\rangle$ be an algebraic system. The element $g \in A$ is a **left-**

identity for $\circ$ iff

$$g \circ a = a \text{ for } all\ a \in A$$

The element d is a **right-identity** for $\circ$ iff

$$a \circ d = a \text{ for } all\ a \in A$$

If e is both a left-identity and right-identity for $\circ$, then e is called an **identity** (or a "two-sided" identity) for $\circ$. (An identity is sometimes called a *neutral element* or a *unity element.*)

Familiar systems such as $\langle R,+\rangle$, $\langle P(A),\cup\rangle$, and $\langle Q_+,\cdot\rangle$ have only one identity—0, $\varnothing$, and 1, respectively—but a system may have many left-identities. For a somewhat artificial example, let A be any set and define $a \circ b = b$ for all $a,b \in A$. This is an operation in A, and every element of A is a left-identity (but none is a right-identity if A has more than one element).

Operations in the same system may have different identities, or one operation may have an identity and another operation not have one. Thus in $\langle J,+,\cdot\rangle$, 0 is an identity for addition and 1 is an identity for multiplication; whereas in $\langle J_+,+,\cdot\rangle$, 1 is an identity for multiplication but there is no identity for addition. This example also shows that the property of having an identity is not hereditary (since $0 \notin J_+$). If an operation has an identity and the operation is denoted by $+$, then 0 will be used to denote its identity.

Definition 1-19

Let $\langle A,\circ\rangle$ be an algebraic system which has an identity e for operation $\circ$. If $a \in A$, then the element $b \in A$ is called a **left-inverse** of a iff

$$b \circ a = e$$

The element c is a **right-inverse** of a iff $a \circ c = e$. An element a' which is both a left-inverse and a right-inverse of a is called an **inverse** (or "two-sided" inverse) of a.

If the operation is denoted by $+$, the inverse of a will be denoted by ${}^-a$ or $-a$. In general, a two-sided inverse of a will be denoted by a^{-1}. Questions of uniqueness of identities and inverses will be left to the exercises, but it should be noted that in the real number system

$\langle R,+,\cdot\rangle$, the additive inverse of a number x is the usual negative $-x$ and the multiplicative inverse of a number $z \neq 0$ is $1/z$. Inverses do not exist, according to the definition, in $\langle R,-\rangle$ since this system has no identity (only the right-identity 0). The system $\langle R_0,+\rangle$ has an identity, but only the identity (0) has an inverse (also 0). Also the property of having inverses is not hereditary since every element has an inverse in $\langle R,+\rangle$, only one does in $\langle R_0,+\rangle$, and none in $\langle R_+,+\rangle$, each of which is an algebraic system.

Exercises 1-56 to 1-60

1-56. Discuss the existence of left-identities and right-identities in:
a. $\langle P(A),\sim\rangle$ b. $\langle P(A),\cap\rangle$
c. $\langle P(A),\cup\rangle$ d. $\langle R_+,\div\rangle$

1-57. Prove: Let $\langle A,\circ\rangle$ be a system with left-identity g and right-identity d. Then $g = d$. As corollaries, show that (a) a two-sided identity is unique, and (b) if $\circ$ is commutative, then $\langle A,\circ\rangle$ has at most one left-identity.

1-58. Prove: Let $\langle A,\circ\rangle$ be a system with identity e in which $\circ$ is associative. If b is a left-inverse for $a \in A$ and c a right-inverse for a, then $b = c$. Consider corollaries similar to those in Exercise 1-56.

1-59. Consider the existence of inverses in $\langle P(A),\cup\rangle$, $\langle P(A),\cap\rangle$, and $\langle R_+,\cdot\rangle$.

1-60. If $\langle A,\circ\rangle$ is an associative system with identity e, and if a^{-1} exists, then $(a^{-1})^{-1} = a$. If a^{-1} and b^{-1} exist, then $(a \circ b)^{-1}$ exists and equals $b^{-1} \circ a^{-1}$.

▲

When considering an algebraic system $\langle A,\circ\rangle$ in which A is a finite set with n elements, one has a particularly complete way of displaying basic information about the system. This display is called an *operation table.* An operation table for $\circ$ is an $n \times n$ square array (matrix) of the elements of A with each element of A listed once at the head of each column and once at the head of each row (preferably in the same order both times). A sample table for a system $\langle\{a,b,c\},\circ\rangle$ is given in

Table 1-1. A different operation $*$ in $\{a,b,c\}$ is given by Table 1-2. Subject to the convention that $x \circ y$ is the entry in the table at the intersection of the x row and y column, the tables actually define the operations $\circ$ and $*$. Conversely, and more usefully, once an operation has been defined by other means in a finite set, its table can be constructed. Thus if ordinary multiplication is the operation in $\{-1,0,1\}$, the table for $\langle\{-1,0,1\},\cdot\rangle$ is Table 1-3.

Table 1-1

$\circ$	a	b	c
a	a	b	c
b	c	b	a
c	c	c	c

Table 1-2

$*$	a	b	c
a	a	b	c
b	b	c	a
c	c	a	b

Table 1-3

$\cdot$	-1	0	1
-1	1	0	-1
0	0	0	0
1	-1	0	1

From such tables many of the properties of operations can be discovered. The operation is commutative iff the table is symmetric in its diagonal from upper left to lower right (Tables 1-2 and 1-3). The operation has an identity iff the row and column for some fixed element reproduce the row and column headings (Table 1-2, a; Table 1-3, 1; Table 1-1, none, although a is a left-identity). Inverses of elements can be found by looking for the identity in the body of the table. (In Table 1-2, $a^{-1} = a$; and $b^{-1} = c$ since $b * c = a$, the identity; and $c^{-1} = b$. In Table 1-3, $1^{-1} = 1$, $(-1)^{-1} = -1$, and 0 has no inverse.)

If an operation has an identity, it will usually be convenient to list it first among the row and column headings—in which case one might even omit the headings. In this way Table 1-3 might be changed from

$\cdot$	-1	0	1
-1	1	0	-1
0	0	0	0
1	-1	0	1

to

$\cdot$	1	0	-1
1	1	0	-1
0	0	0	0
-1	-1	0	1

or to

·			
	1	0	−1
	0	0	0
	−1	0	1

where the first row and column serve also as headings.

Exercises 1-61 to 1-65

1-61. Consider the following table:

·	0	1	2	3	4	5
0	0	0	0	0	0	0
1	0	1	2	3	4	5
2	0	2	4	0	2	4
3	0	3	0	3	0	3
4	0	4	2	0	4	2
5	0	5	4	3	2	1

Is the operation · commutative? Is there an identity? What elements have inverses?

1-62 to 1-65. (***An algebraic system of mappings***) Consider the six mappings a, b, c, d, e, and f defined by

$$xa = x \qquad xd = \frac{1}{1-x}$$

$$xb = \frac{1}{x} \qquad xe = 1 - \frac{1}{x}$$

$$xc = 1 - x \qquad xf = \frac{x}{x-1}$$

1-62. If the domain of each of these mappings is taken to be $R \sim \{0,1\} = D$, the set of all real numbers except 0 and 1, then each of these mappings is a one-to-one mapping of D onto itself. Draw the graphs and verify this geometrically.

☆***1-63.*** Let the operation, composition of mappings, be denoted by juxtaposition in $F = \{a,b,c,d,e,f\}$. Construct the operation table for $\langle F, \ \rangle$—thus verifying that composition is an operation in F.

1-64. Find the identity. Find the inverse of each element. Show that the operation is not commutative.

1-65. Construct the table for the "subsystem" $\langle\{a,d,e\}, \ \rangle$. Show that this system is commutative.

▲

One of the most important ideas in all of modern abstract algebra is that of a mapping between algebraic systems which preserves operations. In discussing these, we are introducing a combination of two earlier ideas—mappings and operations. Some significant uses and properties of such mappings will arise later in this book—notably in Chapters 3, 4, and 6—and the reader who studies advanced topics in algebra will find the subject saturated with references to operation-preserving mappings (also called *homomorphisms*). The definition will be given for systems with one operation.

Definition 1-20

Let $\langle A,\circ\rangle$ and $\langle B,*\rangle$ be algebraic systems. A mapping $\alpha : A \to B$ is called **operation-preserving** (which can be abbreviated to OP) iff

$$(x \circ y)\alpha = (x\alpha) * (y\alpha) \text{ for all } x,y \in A$$

It is important to notice that the equation is an equality of elements of B, not of A. The equation says simply that the answer is the same whether mapping or operating is done first. More precisely, if α is an OP mapping and two elements of A are given, then by operating on those two elements by $\circ$ and mapping the "$\circ$ product" to B, one obtains the same element of B as is obtained by mapping the elements separately to B and forming the "$*$ product" of the images.

A good example of such a mapping is the logarithm mapping. The equation

$$\log (x \cdot y) = \log x + \log y$$

is an equation of the OP form. The respective systems involved are $\langle R_+, \cdot \rangle$ and $\langle R,+\rangle$. The logarithm mapping is also one-to-one and onto; however, these properties are not assumed to hold for all OP mappings. Theorem 1-3 will say that, for most purposes, an OP mapping might as well be onto—in the sense that the part of B which is of interest is rn (α).

Theorem 1-3

Let α be an OP mapping from the system $\langle A,\circ\rangle$ to $\langle B,*\rangle$. Then $\langle \text{rn}\ (\alpha),*\rangle$ is an algebraic system. [That is, if $u,v \in \text{rn}\ (\alpha)$, then $u * v \in \text{rn}\ (\alpha)$.]

PROOF. Let $u,v \in \text{rn}\ (\alpha)$. Then there are elements x and y in A such that $x\alpha = u$ and $y\alpha = v$. Thus

$$u * v = (x\alpha) * (y\alpha) = (x \circ y)\alpha$$

But $x \circ y$ is in A, which means that its image $(x \circ y)\alpha$ is in rn (α). Since $(x \circ y)\alpha = u * v$, the theorem is proved. ||

It remains to be seen in what sense such mappings actually preserve operations. Actually, essentially *all* of the properties of the operation $\circ$ in A are also properties of the operation $*$ in rn (α).

Rather than refer to rn (α) in what follows, it will be easier to make the simplification suggested by Theorem 1-3 of assuming that α is a mapping onto B, so that rn $(\alpha) = B$.

Theorem 1-4

Let α be an OP mapping of $\langle A,\circ\rangle$ onto $\langle B,*\rangle$. If $\circ$ is commutative in A, then $*$ is commutative in B.

DISCUSSION. The design of the proof is dictated by the fact that we want to prove something about elements of B, so the place to start is with elements of B. We shall trace their ancestry back to A and use the property assumed to hold for elements of A—then return to B. This is the basic structure of all such proofs. (Compare Theorem 1-3.)

PROOF. Let $u,v \in B$. Since α maps A onto B, there exist $x,y \in A$ such that $x\alpha = u$ and $y\alpha = v$. Using, twice, the fact that α is an OP mapping sandwiched around the invoking of the commutativity of $\circ$, we calculate that

$$u * v = (x\alpha) * (y\alpha) = (x \circ y)\alpha = (y \circ x)\alpha = (y\alpha) * (x\alpha) = v * u \quad \|$$

Theorem 1-5

The properties of associativity, existence of identities, and existence of inverses are all preserved by an OP mapping.

Proofs of the parts of this theorem and some more precise information about the corresponding identities and inverses in the range of the mapping are deferred to the exercises.

Exercises 1-66 to 1-71

1-66. Show that the mapping $x \to 2^x$ is an OP mapping of $\langle R,+\rangle$ to $\langle R_+, \cdot\rangle$.

1-67. Consider $\langle A,\circ\rangle$ and $\langle B,*\rangle$ defined by the tables

A: $\circ$	a	b	c	d
a	a	b	c	d
b	b	a	d	c
c	c	d	a	b
d	d	c	b	a

and

B: $*$	0	1
0	0	1
1	1	0

Let $\beta : A \to B$ be defined by $a\beta = b\beta = 0$ and $c\beta = d\beta = 1$. Show that β is an OP mapping.

1-68 to 1-70. (***Proof of Theorem 1-5***) Let α be an OP mapping of $\langle A,\circ\rangle$ onto $\langle B,*\rangle$. Prove the following:

1-68. If $\circ$ is associative in A, then $*$ is associative in B.

1-69. If e is an identity in A for $\circ$, then $e\alpha$ is an identity in B for $*$.

1-70. If a^{-1} is an inverse of $a \in A$, then $a^{-1}\alpha$ is an inverse of $a\alpha$ in B. [That is, $a^{-1}\alpha = (a\alpha)^{-1}$.]

1-71. Let $\langle A,+,\cdot\rangle$ and $\langle B,\circ,*\rangle$ be algebraic systems and let α be an OP mapping from $\langle A,+\rangle$ onto $\langle B,\circ\rangle$ and also an OP mapping from $\langle A,\cdot\rangle$ onto $\langle B,*\rangle$. Show: If $\cdot$ is distributive over $+$ in A, then $*$ is distributive over $\circ$ in B.

1-F THE NUMBER SYSTEMS

The algebraic properties of the number systems are the principal subject of much of elementary mathematics. They are intuitively familiar to most educated people—including many who use them only for such mundane purposes as preparing fractional parts of a recipe or computing a bank balance. It is not the purpose of this section to develop the number systems from a minimal set of postulates and definitions. We shall be content to list here certain properties of the usual number systems—properties we shall take to be basic. The consequences of these properties and, therefore, other characteristics of the systems having these properties will be revealed through the medium of the general theories developed in the rest of the book. Of the number systems mentioned here, only the nonnegative integers will receive an expanded treatment.

The nonnegative integers. The system $\langle J_0,+,\cdot\rangle$ of nonnegative integers consists of the set $J_0 = \{0,1,2, \ldots\}$ and two operations, addition (+) and multiplication (·), such that

S: (Closure) Addition and multiplication are operations in J_0.
A: (Associativity) Both operations are associative.
C: (Commutativity) Both operations are commutative.
D: (Distributivity) Multiplication is distributive over addition.
I: (Identities) The number 0 is the identity for addition, and the number 1 is the identity for multiplication.
N: (Induction) If M is a subset of J_0 such that (1) $0 \in M$ and (2) if $m \in M$, then $m + 1 \in M$, *then* $J_0 = M$.

Of these properties only the most subtle one N, the *principle of mathematical induction*, was not considered in the last section. This property (axiom) is important in that it provides a way to prove some theorems about J_0. In fact, it applies to subsets of J_0 such as

$$\{n \in J_0 \mid n \geq k\} \text{ for fixed } k \in J_0$$

in the form: "If $M \subset J_0$, if $k \in M$, and if $m + 1 \in M$ whenever $m \in M$ for all $m \geq k$, then $\{n \in J_0 \mid n \geq k\} \subset M$." In the most usual sort of application of induction, one attempts to prove about numbers a proposition of the form: "If $n \in J_0$ (or if $n \geq k$), then $P(n)$ (a statement about n) is true." The machinery for induction is established by letting $M = \{n \mid n \in J_0 \text{ and } P(n) \text{ is true}\}$. Then find the smallest member k of J_0 for which $P(k)$ is true. ($k = 0$, 1, or 2 in most applications.) Finally assume that $P(m)$ is true for some arbitrary $m \geq k$ (the *induction hypothesis*), and show that this implies that $P(m + 1)$ is true. That is, show that $k \in M$ and that, whenever $m \in M$, then $m + 1 \in M$. Thus $P(n)$ is true for all $n \geq k$. The intuitive idea of induction is that, if there is a line of people in which the first person is wearing a hat and every person behind a person wearing a hat also is wearing a hat, then everybody has on a hat.

As an example of an inductive proof, let us show that

$$2^0 + 2^1 + 2^2 + \cdots + 2^n = 2^{n+1} - 1$$

for every $n \in J_0$. To do this, we first note that $1 = 2^0 = 2^{0+1} - 1$. If we let

$$M = \{n \mid 2^0 + 2^1 + \cdots + 2^n = 2^{n+1} - 1\}$$

we know that $0 \in M$. Next we assume that $m \in M$—that is, that $2^0 + 2^1 + \cdots + 2^m = 2^{m+1} - 1$. We want to show that $m + 1 \in M$—that $2^0 + 2^1 + \cdots + 2^m + 2^{m+1} = 2^{(m+1)+1} - 1$. Our calculation proceeds as follows:

$$\begin{aligned} 2^0 + 2^1 + \cdots + 2^m + 2^{m+1} & \\ &= (2^0 + 2^1 + \cdots + 2^m) + 2^{m+1} && \text{Associativity} \\ &= (2^{m+1} - 1) + 2^{m+1} && \text{Since } m \in M \\ &= 2 \cdot 2^{m+1} - 1 \\ &= 2^{(m+1)+1} - 1 \end{aligned}$$

Thus $m + 1 \in M$ whenever $m \in M$. Since $0 \in M$, we conclude by property N that $M = J_0 = \{0,1,2,3, \ldots\}$.

A weaker form of the principle of mathematical induction can be formulated as follows:

N′: If M is a subset of J_0 such that (1) $0 \in M$ and (2) if all integers n such that $0 \leq n \leq m$ are in M, then $m + 1 \in M$, *then* $J_0 = M$.

Apparently N′ assumes more than N, namely that not just $m \in M$, but m and all nonnegative integers less than m are in M, will force $m + 1$ to be in M. In both the formulations N and N′, the heart of the proof is to show that the "if" part of (2) implies the "then" part of (2). [But don't forget to prove the (1) part.]

It is relatively easy to see that properties N and N′ are logically equivalent.

Some further properties of J_0 appear in Section 4-D.

The integers. The system of integers is $\langle J,+,\cdot\rangle$ with

$$J = \{0,1,-1,2,-2, \ldots\}$$

satisfying, in addition to properties S, A, C, D, and I, the property

V^+: (Inverse) Each element of J has an inverse, called its *negative* for addition.

The fact that the word "negative" is used in two senses occasionally causes some trouble. It is common practice to speak of "negative numbers" and of "the negative of a number." The trouble comes largely from the fact that the negative of a number may not be a negative number. In fact, the negative of a negative number is a positive number. The negative of -7 is $-(-7) = 7$, for example. This confusion can occur in the definition and use of absolute value.

Definition 1-21

The **absolute value** of an integer n is n if $n \geq 0$ and $-n$ if $n < 0$. The absolute value of n is denoted by $|n|$.

Thus $|0| = 0$, $|6| = 6$, and $|-4| = -(-4) = 4$.

Theorem 1-6 (**The division algorithm**)

If n and d are integers and $d \neq 0$, then there exist unique integers q and r such that $0 \leq r < |d|$ and $n = dq + r$. The number q is called the *quotient* and the number r the *remainder* on dividing n by d.

We shall not prove this theorem here. (Similar results will appear later.) In effect, what Theorem 1-6 says is that, if one divides n by d, he will get a quotient q and a remainder r, which are unique if he insists that the remainder is between 0 and $|d|$ but not equal to $|d|$. Dividing 17 by 3 gives $17 = 3 \cdot 5 + 2$. If $n = 12$ and $d = 4$, the result is $12 = 4 \cdot 3 + 0$. If $n = -35$ and $d = 8$, then $-35 = 8 \cdot (-5) + 5$ [not $-35 = 8 \cdot (-4) - 3$ because $r \geq 0$]. If $n = 17$ and $d = -5$, then $17 = (-5) \cdot (-3) + 2$. If $n = -27$ and $d = -4$, then $-27 = (-4) \cdot 7 + 1$.

The integers will be completely characterized from an algebraic point of view in Chapter 4.

The rational numbers. The system of rational numbers is $\langle Q, +, \cdot \rangle$ satisfying, in addition to properties S, A, C, D, I, and V^+, the property

$V^\cdot$: (Inverse) If $a \in Q$ and $a \neq 0$, then a has an inverse for multiplication.

The rational numbers will also be given a complete algebraic characterization in Chapter 4.

The real numbers. The system of real numbers $\langle R, +, \cdot \rangle$ satisfies all the properties S, A, C, D, I, V^+, and $V^\cdot$ listed above but has an additional property, called *completeness*, which distinguishes it from the system of rational numbers. Most recent books in analysis (calculus, advanced calculus, real variables) discuss completeness in one of its several forms in detail. It is this idea which makes $\sqrt{2} \in R$, whereas $\sqrt{2} \notin Q$. The interested reader is referred to an analysis book for the details of this idea. See the references at the end of this chapter for suggested sources.

Order. Each of the systems mentioned above is ordered by the relation $<$, which has the following properties:

O_1: Given two numbers a and b, exactly one of the statements $a = b$, $a < b$, $b < a$ is true.
O_2: If $a < b$ and $b < c$, then $a < c$.
O_3: If $a < b$ and c is any number, then $a + c < b + c$.
O_4: If $a < b$ and $0 < c$, then $a \cdot c < b \cdot c$.

In Chapter 4 the properties of order will be introduced using a different set of axioms, and the properties O_1 to O_4 above will become theorems.

The complex numbers. An example of a system satisfying all of the properties of the real numbers except the order properties is the system of complex numbers. A sophisticated view of the complex numbers will be presented at the end of Chapter 5, and they will be described now merely as a system $\langle C,+,\cdot\rangle$ whose elements have the form $a + bi$ with $a,b \in R$ and behave algebraically like real numbers subject to the rule that $i^2 = -1$.

Exercises 1-72 to 1-87

1-72. Prove for every $n \in J_+$:

a. $1 + 2 + 3 + \cdots + n = n(n + 1)/2$

b. $1^3 + 2^3 + 3^3 + \cdots + n^3 = n^2(n + 1)^2/4$

[Note that (a) and (b) imply the rather interesting property $(1 + 2 + \cdots + n)^2 = 1^3 + 2^3 + \cdots + n^3$. The reader may be interested in developing a noninductive proof of this result.]

c. $1 + r + r^2 + \cdots + r^{n-1} = \dfrac{1 - r^n}{1 - r}$, $r \in R$ and $r \neq 1$

d. $1 \cdot 2 + 2 \cdot 3 + 3 \cdot 4 + \cdots + n(n + 1) = \dfrac{n(n + 1)(n + 2)}{3}$

e. $\dfrac{1}{1 \cdot 2} + \dfrac{1}{2 \cdot 3} + \dfrac{1}{3 \cdot 4} + \cdots + \dfrac{1}{n(n + 1)} = \dfrac{n}{n + 1}$

1-73 to 1-75. (***Well-ordering***) A set W is *well-ordered* by a relation $<$ iff every nonempty subset of W contains a smallest element (an element b such that $b \leq x$ for every x in the subset).

1-73. Use the given properties of the nonnegative integers (especially N or N′) to prove that J_0 is well-ordered by $<$.

1-74. Show by examples that neither J nor Q_0 is well-ordered by $<$.

☆*1-75.* (Converse of 1-73) Use the well-ordering property of J_0 and the given properties except for N, and prove N. (Thus well-ordering and mathematical induction are logically equivalent.)

☆*1-76.* Prove that, if $a \in J$, then $a \cdot 0 = 0$. List a property of $\langle J, +, \cdot \rangle$ to justify each step.

☆*1-77.* Use the result of Exercise 1-76 to show that $a \cdot (-b) = -(a \cdot b)$ for all $a,b \in J$.

1-78. If $a,b,c \in J$ and $a < b$ and $c < 0$, then $bc < ac$.

1-79. Find the numbers q and r of Theorem 1-6 if

a. $n = 12$, $d = -4$ b. $n = 0$, $d = 7$

c. $n = -49$, $d = 11$ d. $n = 2$, $d = -5$

1-80. Find a fallacy in the following induction argument:

PROPOSITION. In any set of n numbers, all the numbers are equal.

PROOF. Let $M = \{n \mid n \in J_+$ and in all sets of n numbers all the numbers are equal$\}$. Clearly $1 \in M$. Assume $m \in M$. Let $B = \{a_1,a_2, \ldots, a_m,a_{m+1}\}$ be any set of $m + 1$ numbers. Then the subsets $\{a_1,a_2, \ldots ,a_m\}$ and $\{a_2, \ldots ,a_m,a_{m+1}\}$ are both sets of m numbers. By the induction hypothesis all numbers in each set are equal. That is, $a_1 = a_2 = \cdots = a_m$ and $a_2 = \cdots = a_m = a_{m+1}$. Then also

$$a_1 = a_2 = \cdots = a_m = a_{m+1}$$

so all members of B are equal, and $m + 1 \in M$.

1-81 to 1-85. (***Congruent integers***) Define a relation $\equiv_n$, called *congruence modulo n*, in J for a fixed $n \in J_+$ by

$$a \equiv_n b \text{ iff there exists } k \in J \text{ such that } a - b = kn$$

(that is, $a-b$ is divisible by n). Note that k is not necessarily positive.

1-81. List three integers, at least one of which is negative, congruent modulo 7 to each of 2, 5, and 48.

1-82. Show that $\equiv_n$ is an equivalence relation in J for each $n \in J_+$.

1-83. If $a \equiv_n b$ and $c \equiv_n d$, then $a + c \equiv_n b + d$ and $ac \equiv_n bd$.

☆***1-84.*** If E_a and E_b are equivalence classes in J for the relation $\equiv_n$, then an addition and multiplication of classes defined by

$$E_a + E_b = E_{a+b} \qquad \text{and} \qquad E_aE_b = E_{ab}$$

are independent of the choice of elements used to represent the classes. (That is, if $E_c = E_a$, then $E_{c+b} = E_{a+b}$ and $E_{cb} = E_{ab}$.)

1-85. The addition and multiplication of equivalence classes defined in Exercise 1-84 are associative and commutative, and multiplication is distributive over addition.

1-86. Find identities for addition and multiplication of equivalence classes and show that each class has an additive inverse.

1-87. How many distinct equivalence classes are there for the relation $\equiv_n$ in J? Make addition and multiplication tables for the distinct equivalence classes of $\equiv_4$.

Miscellaneous exercises 1-88 to 1-89

1-88. What value should be assigned to the symbols $\bigcup_{t \in \varnothing} A_t$ and $\bigcap_{t \in \varnothing} A_t$ in keeping with Definition 1-3 and Exercise 1-14?

1-89. Let S be a set with n elements.

a. How many (binary) operations can be defined in S?
b. How many of these are commutative?
c. How many operations with identity can be defined in S?
d. In how many of these does each element have an inverse?
e. How many commutative operations with identity can be defined in S?

References for further study

THE THEORY OF SETS

Halmos, P. R.: "Naive Set Theory," D. Van Nostrand Company, Inc., Princeton, N.J., 1960.

COMPLETENESS OF REAL NUMBERS

Boas, Jr., R. P.: "A Primer of Real Functions," John Wiley & Sons, Inc., New York, 1960.

Johnson, R. E., and F. L. Kiokemeister: "Calculus with Analytic Geometry," 3d ed., Chap. 7, Allyn and Bacon, Inc., Boston, Mass., 1964.

Munroe, M. E.: "Introductory Real Analysis," Addison-Wesley Publishing Company, Inc., Reading, Mass., 1965.

2 GROUPS

The first collection of abstract systems to be studied in detail is a collection of one-operation systems. These systems are called groups, and there are many examples that indicate the importance of groups. In particular, groups appear in the usual number systems when one considers just one operation at a time, groups appear in geometry—both synthetic and analytic—and groups appear in analysis as certain systems of mappings.

A group is a relatively simple system in that it involves only one operation and three assumptions about it.

2-A BASIC PROPERTIES

The system $\langle X,+\rangle$, where X is the set of integers, or the rational, real, or complex numbers has the properties of associativity, commutativity, an identity (0) and inverses for each element. It is these properties which are considered when defining a group. The property of commutativity can conveniently be omitted, however, because many other systems can be included as groups if the operation is not taken to be commutative. It is the plan of this section to show also that general ideas of identity and inverses can be obtained from a still more economical set of axioms. The advantage of this will be felt when it becomes necessary to show that a given system is a group. It will be easier to verify that the axioms hold than to verify the existence of identities and inverses.

Definition 2-1

A **group** is an algebraic system $\langle G,\circ\rangle$ such that:

G1: The operation $\circ$ is associative.
G2: There is a left-identity $g \in G$ for $\circ$.
G3: For every $x \in G$ and every left-identity $g \in G$ there exists $x' \in G$ such that $x' \circ x = g$.

Actually there is a fourth assumption implicit in the first line of this definition in the statement, ". . . is an algebraic system . . . ," which implies (Definition 1-15) that the operation $\circ$ satisfies the *closure* property:

G0: If $x,y \in G$, then $x \circ y \in G$.

Among examples of groups are the systems $\langle R,+\rangle$, $\langle Q_+,\cdot\rangle$, $\langle J,+\rangle$, and the system $\langle F,\ \rangle$ of Exercise 1-63. Other examples include $\langle\{0\},+\rangle$ and $\langle\{-1,1\},\cdot\rangle$ which have only one and two elements, respectively. A group must have at least one element by Axiom G2. Notice that the identities in each example (0, 1, 0, a, 0, and 1, respectively) are two-sided and that all inverses in these examples are also two-sided—even in the example $\langle F,\ \rangle$ where the operation is not commutative. It is our first objective to show that this is always the case.

Theorem 2-1

If $\langle G,\circ\rangle$ is a group with left-identity g, then the following are true:

1. If $a,b,c \in G$ and $a \circ b = a \circ c$, then $b = c$.
2. If $a \in G$, then $a \circ g = a$.
3. The element g is a unique two-sided identity for $\circ$.

PROOF. *1.* Assume $a \circ b = a \circ c$. By G3, there exists a' such that $a' \circ a = g$, a left-identity. Thus

$$a' \circ (a \circ b) = a' \circ (a \circ c) \qquad \text{Definition 1-14}$$
$$(a' \circ a) \circ b = (a' \circ a) \circ c \qquad \text{G1}$$
$$g \circ b = g \circ c \qquad \text{G3}$$
$$b = c \qquad \text{G2}$$

2. By G3, there exists an a' in G such that $a' \circ a = g$. Therefore

$$g \circ g = g \qquad \text{G2}$$
$$(a' \circ a) \circ g = a' \circ a \qquad \text{G3}$$
$$a' \circ (a \circ g) = a' \circ a \qquad \text{G1}$$
$$a \circ g = a \qquad \text{Part 1}$$

3. If g and h are both left-identities for $\circ$, then they are both right-identities by part 2. Then $g \circ h = g$ since h is a right-identity, and $g \circ h = h$ since g is a left-identity. Therefore $g = h$. (This was Exercise 1-57.) ||

Part 1 of this theorem is usually called the *left cancellation law*. Parts 2 and 3 guarantee that a group has exactly one identity. It is also true that each element has exactly one two-sided inverse, and that inverse is the left-inverse guaranteed by G3. This is the content of the following theorem.

Theorem 2-2

Let $\langle G,\circ\rangle$ be a group with identity e. Then

1. If a' is a left-inverse of $a \in G$, then $a \circ a' = e$.

2. If $a,b,c \in G$ and $a \circ c = b \circ c$, then $a = b$.

3. If $a,b \in G$ and $b \circ a = e$, then $b = a'$.

DISCUSSION. Part 2 is, of course, called the *right cancellation law*. Parts 1 and 3, taken together, guarantee that any inverse of an element is two-sided and unique. The proof of this theorem is deferred to the exercises.

In comparing the cancellation laws—Theorem 2-1, part 1, and Theorem 2-2, part 2—notice that nowhere is the possibility of cross-cancellation implied. Thus if $b \circ a = a \circ c$, it is not necessarily true that $b = c$. However, if $\circ$ is commutative (which was not assumed in Definition 2-1), then we would have $b = c$. As an example in the noncommutative case, consider the group $\langle F, \ \rangle$ of Exercise 1-63. There $cd = b = df$, but $c \neq f$. An additional piece of information which can be inferred from the cancellation laws is a property of the table for a group $\langle G,\circ\rangle$ where G is a finite set: namely, every element occurs exactly once in each row and each column of the table. This follows because, if any element x occurs twice in some column and if b is the head of the column and c and d the heads of the rows in which x occurs, then $c \circ b = d \circ b = x$, so $c = d$ by right cancellation. (Left cancellation yields a similar result if some element occurs twice in the same row.)

As mentioned above, to show that a given system is a group, it is necessary to show only that it is a system (closure) and that properties G1, G2, and G3 are satisfied. It should be clear that to show a right-identity and right-inverse would also suffice. (However, right-identity and left-inverse—or vice versa—does not necessarily yield a group.) In other words, it is possible to prove that a system is a group relatively easily. On the other hand, once it is known that a system $\langle G,\circ\rangle$ *is* a group, then each of the following four statements is true.

G0′: The operation $\circ$ is an operation in A (closure).

G1′: The operation $\circ$ is associative in A (associativity).

G2′: There is a unique element $e \in A$ such that $e \circ a = a \circ e = a$ for every $a \in A$ (identity).

G3′: If $a \in A$, then there is a unique element $a^{-1} \in A$ such that $a \circ a^{-1} = a^{-1} \circ a = e$ (inverses).

These four are called the *classical group axioms*, and they are sometimes still used as the definition of a group.

A little more terminology will enable the discussion of items which follow to proceed more smoothly.

Definition 2-2

A group $\langle G, \circ \rangle$ is called a **commutative group,** or an **abelian group,** iff the operation $\circ$ is commutative in G.

Definition 2-3

The **order** of a group $\langle G, \circ \rangle$ is the number of elements of G. If G is a finite set with n elements, then $\langle G, \circ \rangle$ is a *finite group of order n*. If G is an infinite set, then $\langle G, \circ \rangle$ is a *group of infinite order.*

The word "abelian" in Definition 2-2 is used in honor of the Norwegian mathematician Niels Henrik Abel (1802–1829), who studied such groups and made many other significant contributions in the areas of algebra and analysis during his brief life.

A couple of pieces of information about inverses are contained in Exercise 1-60 and are provable from the axioms of a group, as can be seen by checking the assumptions made there. We state these for emphasis as a theorem.

Theorem 2-3

If $\langle G, \circ \rangle$ is a group and if $a,b \in G$, then $(a^{-1})^{-1} = a$ and

$$(a \circ b)^{-1} = b^{-1} \circ a^{-1}$$

Exercises 2-1 to 2-9

2-1. Show that each of the following is an abelian group:

a. $\langle R, + \rangle$ b. $\langle Q_+, \cdot \rangle$

2-2. Show that each of the following is not a group:

a. $\langle J_0, + \rangle$ b. $\langle Q, \cdot \rangle$

c. $\langle R, - \rangle$ d. $\langle P(A), \cup \rangle$

e. $\langle \{0,1,-1\}, + \rangle$

2-3. Which of the following are groups?

a. $\langle P(A), \cap \rangle$

b. $\langle \{0,1,-1\}, \cdot \rangle$

c. $\langle \{a,b\}, \quad \rangle$, where $\{a,b\}$ is a subset of the set F of Exercise 1-63
d. $\langle \{a,d\}, \quad \rangle$ of the same exercise
e. $\langle R \sim \{0\}, \cdot \rangle$

2-4. Let $G = \{(a,b) \mid a,b \in R \text{ and } a \neq 0\}$. Let

$$(a,b) * (c,d) = (ac, bc + d)$$

in G. Show $\langle G,* \rangle$ is a group.

☆**2-5.** Let $\langle G,\circ \rangle$ be a group in which $a \circ a = e$, the identity, for every $a \in G$. Prove that $\circ$ is commutative. Note that $a \circ a = e$ is another way of saying $a = a^{-1}$.

☆*2-6.* Prove Theorem 2-2.

2-7. We know that b is the identity in a group $\langle G,\circ \rangle$ if $a \circ b = a$ for every $a \in G$. Show that b is the identity if $a \circ b = a$ for *some* $a \in G$.

2-8. An element d in an algebraic system is called *idempotent* iff $d \circ d = d$. Show that the identity is the only idempotent element of a group.

2-9. Consider the set $R \sim \{1\}$ and the operation $\circ$ defined in terms of ordinary addition and multiplication by

$$a \circ b = a + b - ab$$

Show that $\langle R \sim \{1\}, \circ \rangle$ is a commutative group. (Show first that if $a \neq 1$ and $b \neq 1$, then $a \circ b \neq 1$.)

▲

Although a group is a relatively weak system in that few assumptions are made, it is the case that most of the more cherished properties of the algebra of real numbers are true in any group. The cancellation laws are examples of this. The solvability of "linear" equations is another example, and this is what Theorem 2-4 is about.

Theorem 2-4

Let $\langle G,\circ \rangle$ be a group and let $a,b \in G$. Then there exist unique elements x and y in G such that $a \circ x = b$ and $y \circ a = b$.

DISCUSSION. What this theorem says is that if $a \circ x = b$ is an equation in $\langle G,\circ\rangle$ with "unknown" x, then there is an element of G ($x = a^{-1} \circ b$) which satisfies the equation. The proof of the uniqueness follows a standard pattern for uniqueness proofs (see Theorem 2-1) of assuming two solutions and showing them to be the same element.

PROOF. The proof will be given for $a \circ x = b$. The case $y \circ a = b$ is similar. Multiplying both sides of $a \circ x = b$ by a^{-1} on the left yields $x = a^{-1} \circ b$, which is a solution since

$$a \circ x = a \circ (a^{-1} \circ b) = (a \circ a^{-1}) \circ b = e \circ b = b$$

If there is more than one solution to $a \circ x = b$, then there exist $u,v \in G$ such that $a \circ u = b$ and $a \circ v = b$. But then $a \circ u = a \circ v$, so $u = v$ by the left cancellation law. ||

Exercises 2-10 to 2-16

2-10. Prove the assertions of Theorem 2-4 relating to the equation $y \circ a = b$.

2-11. In the group of Exercise 1-63 solve for y:
a. $yd = c$ b. $dy = c$ c. $byd = e$ d. $cydf = bc$

2-12. In the group of Exercise 2-4, solve for x:
a. $(2,1) * x = (1,0)$ b. $x * (5,2) = (15,17)$
c. $x * (1,1) = (1/2,\pi)$ d. $(1,1) * x = (1/2,\pi)$

2-13. In the group of Exercise 2-9 solve for y:
a. $4 \circ y = 1/3$ b. $y \circ 2/7 = -4$ c. $2 \circ (y \circ 2) = 4$

2-14. In the integers J, let $m \vee n = m + n + 1$. Show that $\langle J, \vee\rangle$ is a group, and solve $50 \vee x = 100$.

☆*2-15.* Let $\langle A,*\rangle$ be an algebraic system in which $*$ is associative, the left and right cancellation laws hold, and A is a finite set. Show that $\langle A,*\rangle$ is a group. Give an example to show that $\langle A,*\rangle$ may not be a group if A is an infinite set.

2-16. Let $\langle B,\circ\rangle$ be an algebraic system in which $\circ$ is associative and the equations $a \circ x = b$ and $y \circ a = b$ have solutions for all $a,b \in B$. Show that $\langle B,\circ\rangle$ is a group.

2-B SUBGROUPS

Notations and descriptions have been fairly precise up to this point. We now relax our formality a little by permitting ourselves the luxury of referring to "the group G" instead of "the group $\langle G,\circ\rangle$" when no confusion over the operation can arise, and we shall use juxtaposition more frequently than we have done previously to indicate an operation.

Some groups (all groups with more than one element) have smaller groups in them, and a great deal can be learned about the original group by studying these smaller groups, which are called *subgroups*, as Chapter 3 will show.

Some examples of subgroups are $\langle Q,+\rangle$ as a subgroup of $\langle R,+\rangle$, the system $\langle\{-1,1\},\cdot\rangle$ as a subgroup of $\langle Q \sim \{0\},\cdot\rangle$, and the system of Exercise 1-65.

Definition 2-4

Let $\langle G,\circ\rangle$ be a group and H a subset of G. Then H is a **subgroup** of G iff $\langle H,\circ\rangle$ is a group.

Note that H is a subgroup iff H is a group with respect to the operation of G. The definition requires using the same operation. To prove that H is a subgroup of G, one must show that $\langle H,\circ\rangle$ is an algebraic system in its own right (that $a,b \in H$ implies $a \circ b \in H$), that H contains a left (right) identity for $\circ$, and that every element of H has a left (right) inverse in H. It is not necessary to show associativity since $\circ$ is already assumed to be associative for all elements of G, and all elements of H are in G. According to the definition, G and $\{e\}$ are always subgroups of G. Any other subgroups are called *proper* subgroups.

One question that should arise and be settled in connection with identities and inverses is that of whether or not new elements can be found to take over these functions in H if the elements which originally performed these functions in G are not in H. The answer is that no substitution is possible.

Theorem 2-5

Let H be a subgroup of G. Then the identity of H is the identity e of G and the inverse of an element $x \in H$ is its inverse x^{-1} in G.

PROOF. If there is an element $i \in H$ such that $i \circ a = a$ for all $a \in H$, then we have $e \circ a = a$ in G and $i \circ a = a$ in G also. Therefore $i = e$ by right cancellation.

The proof for inverses is deferred to the exercises. ||

The next theorem is the most basic result about subgroups. It offers an alternative way of showing that a subset of a group is a subgroup.

Theorem 2-6

Let $\langle G,\circ\rangle$ be a group. Then a subset H of G is a subgroup of G if, and only if, $H \neq \emptyset$ and for all $a,b \in H$, $a \circ b^{-1} \in H$.

DISCUSSION. This theorem says that we can show that $H \subset G$ is a subgroup of G by showing two things: that H is nonempty, and that $a \circ b^{-1}$ is in H whenever a and b are in H. This is most convenient when we don't have to show separately that b^{-1} is in H whenever b is in H. Notice that $\circ$ is associative in H because $\circ$ is associative in G, so three things must be shown—identity, inverses, and closure. We show them in that order.

PROOF. Since this is an "if-and-only-if" theorem, there are actually two theorems to prove. To show, "If H is a subgroup of G, then $H \neq \emptyset$ and $a \circ b^{-1} \in H$ for all $a,b \in H$," is easy because if H is a subgroup of G, then H is a group. So H must contain an element (the identity), the inverse of every $b \in H$, and the "$\circ$ product" of any two elements a and b^{-1} in H.

The converse, "If $H \neq \emptyset$ and $a \circ b^{-1} \in H$ whenever $a,b \in H$, then H is a subgroup of G," can be shown as follows:

1. Since $H \neq \emptyset$, there is an element $a \in H$. Apply the hypothesis to the pair $a,a \in H$ and thus verify that $a \circ a^{-1} = e \in H$.
2. Apply the hypothesis to the pair $e,a \in H$ to see that

$$e \circ a^{-1} = a^{-1} \in H$$

Thus the inverse of each element of H is in H.

3. Finally, let $a,b \in H$. We know by part 2 that $b^{-1} \in H$. Apply the hypothesis to the pair $a,b^{-1} \in H$ to get $a \circ (b^{-1})^{-1} = a \circ b \in H$.

Thus the product of two elements of H is in H. This proves that $\langle H,\circ\rangle$ is a group. ||

In applying Theorem 2-6 to show that a subset H of G is a subgroup of G, the "$H \neq \varnothing$" part is usually done by showing that the identity e is in H. In this case, Theorem 2-6 is actually easier to use than the definition only if it is not necessary to prove that $b^{-1} \in H$ before proving $a \circ b^{-1} \in H$. If one must show $b^{-1} \in H$ first, then he has a choice of showing $a \circ b^{-1} \in H$ (Theorem 2-6) or $a \circ b \in H$ (Definition 2-1).

The hypotheses of Theorem 2-6 can be weakened slightly in looking for finite subgroups.

Theorem 2-7

If K is a finite nonempty subset of a group G such that $ab \in K$ for all $a,b \in K$, then K is a subgroup of G.

DISCUSSION. Note that G is not necessarily a finite group. It is only the subset K which must be finite in order for this theorem to apply. (Of course, if G is finite, then all its subsets are also finite.) The method of proof here is to consider a complete list of products of a certain kind in K and use the finiteness of K to show that the identity and inverses are in K.

PROOF. Let $K = \{a_1,a_2, \ldots ,a_n\}$. Consider all n products of the form a_1a_i ($i = 1,2, \ldots ,n$). If $a_1a_i = a_1a_j$, then $a_i = a_j$ by left cancellation in G, so the elements a_1a_i are all different ($i = 1,2, \ldots ,n$). Since these are all in K by hypothesis and since there are n of them, $\{a_1a_1,a_1a_2,a_1a_3, \ldots ,a_1a_n\}$ must be a list of the n elements of K in some order. In particular, $a_1a_r = a_1$ for some r; so $a_r = e$, the identity of G, by Exercise 2-7; thus $e \in K$. Also $a_1a_s = a_r = e$ for some s; so $a_s = a_1^{-1}$. The inverses of the other elements a_j can be found by finding the a_i such that $a_ja_i = a_r = e$. Thus K is a subgroup of G. ||

Applications of these theorems will be found among the following exercises.

Exercises 2-17 to 2-26

2-17. Is the converse of Theorem 2-7: "If K is a finite subgroup of G, then $K \neq \varnothing$ and $ab \in K$ for all $a,b \in K$," true?

2-18. If H and K are subgroups of G, then $H \cap K$ is a subgroup of G. If H_t is a subgroup of G for all $t \in I$, then $\bigcap_{t \in I} H_t$ is a subgroup of G.

2-19. Find six subgroups of the group of Exercise 1-63. Use Theorem 2-7.

2-20. Find four subgroups of $\langle J,+\rangle$. Can you describe *all* subgroups of $\langle J,+\rangle$?

2-21. Find four subgroups of $\langle Q_+, \cdot\rangle$.

2-22. Is $\langle Q \sim Q_0, \cdot\rangle$ a subgroup of $\langle Q \sim \{0\}, \cdot\rangle$?

2-23. Complete the proof of Theorem 2-5.

2-24. Let $\langle G,\circ\rangle$ be a group and $a \in G$. Consider

$$N_a = \{x \mid x \circ a = a \circ x\}$$

the set of all elements of G that commute with a. Show that N_a is a subgroup of G. (If possible, use Theorem 2-6 and don't show directly that $y^{-1} \in N_a$ if $y \in N_a$.)

2-25. Use Exercises 2-18 and 2-24 to show that

$$C = \{x \mid xa = ax \text{ for all } a \in G\}$$

is a subgroup of a group $\langle G, \ \rangle$. The subgroup C is called the *center* of G.

2-26. Prove Exercise 2-25 directly by using Theorem 2-6.

2-C EXAMPLES OF GROUPS

Several examples of groups, most of them taken from the number systems, have already been mentioned. This section introduces a variety of other examples from various parts of mathematics.

Symmetric difference of sets. The symmetric difference operation Δ on sets was defined in Exercise 1-15. The important result of that exercise is that this operation, unlike $\cup$ and $\cap$, makes a group out of the subsets of a given universal set. Recall that $P(X)$ denotes the set whose elements are the subsets of X.

Definition 2-5

Let A and B be subsets of a universal set X. The **symmetric difference** of A and B is the set $A \Delta B$ defined by

$$A \Delta B = (A \sim B) \cup (B \sim A) = (A \cup B) \sim (A \cap B)$$

(Figure 2-1).

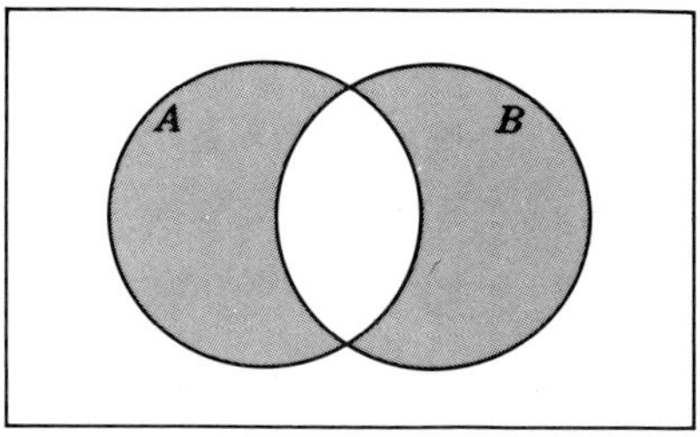

$A\Delta B$

Figure 2-1 $A \Delta B$.

Theorem 2-8

The system $\langle P(X),\Delta\rangle$ is an abelian group. The identity is $\varnothing$ and the inverse of $A \in P(X)$ is A itself.

The proof of this theorem is elementary, except for the rather messy calculation of the associative property of Δ, and is deferred to the exercises. One interesting feature of this group is that each element is its own inverse. (Actually this fact implies that Δ is commutative. See Exercise 2-5.) Also, if X contains n elements, $\langle P(X),\Delta\rangle$ will be a group with 2^n elements. It is possible to show that a finite group with each element its own inverse must have 2^n elements for some n, so a finite group with order different from a power of 2 must contain elements which are not their own inverses.

Exercises 2-27 to 2-30

☆***2-27.*** Make a table for $\langle P(X),\Delta\rangle$ if $X = \{a,b\}$; if $X = \{1,2,3\}$. (Avoid the tedium of writing braces in the table by writing "23" instead of "{2,3}," "2" instead of "{2}," etc.)

2-28. Prove Theorem 2-8 by proving

a. $A \Delta B = B \Delta A$
b. $(A \Delta B) \Delta C = A \Delta (B \Delta C)$
c. $A \Delta \varnothing = A$
d. $A \Delta A = \varnothing$

(You may already have done some of these in Exercise 1-15.)

2-29. Let G be a group in which $a = a^{-1}$ for every $a \in G$. Let $x = y$ be an equation in G, where x and y are products of elements of G. Show that any element occurring twice in the equation may be cancelled. (Use Exercise 2-5.)

2-30. Let G be a finite group in which $a = a^{-1}$ for every $a \in G$. Show that the order of G is 2^n for some $n \in J_0$.

Geometric symmetries. To develop the next examples, consider first a rectangle which is not a square. There are four rigid transforma-

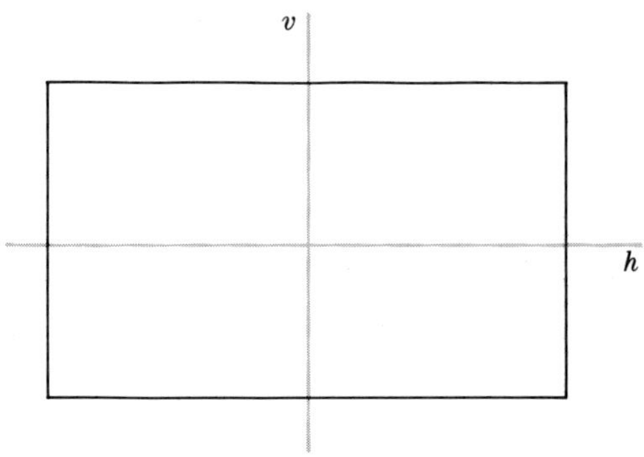

***Figure 2-2** Axes of symmetry of a rectangle.*

tions of the rectangle onto itself; that is, there are four ways of picking up the rectangle and placing it back in the space it originally occupied. These are I the identity (no change), R rotation of 180° in the plane of the rectangle about the center of the rectangle, H rotation of 180° about a horizontal axis h (which turns the rectangle over), V rotation of 180° about a vertical axis v. (See Figure 2-2.) If the product of two of these transformations of the rectangle is defined to be their composition (first followed by second), then the following table is obtained.

	I	R	H	V
I	I	R	H	V
R	R	I	V	H
H	H	V	I	R
V	V	H	R	I

For example, the statement HV = R means that the acts of first flipping the rectangle over horizontally and then flipping it over

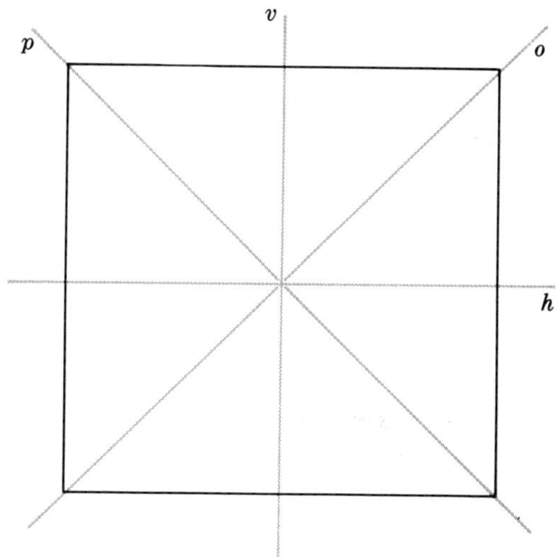

***Figure 2-3** Axes of symmetry of a square.*

vertically result in the same final position of the rectangle as would result from the single act of rotating it 180°. Since the operation in this group is composition of mappings, it is associative by Theorem 1-2. The other group properties follow from the table.

A more complicated example of the same type is obtained by taking the geometric figure to be a square. Now there will be four rotations in the plane and four rotations out of the plane which map the square rigidly onto itself. Let the rotations I, R, R^2, and R^3 be, respectively, the counterclockwise rotations through 0°, 90°, 180°, and 270° in the plane of the rectangle. Let H, V, P, and O be the rotations of the square in the axes h, v, p, and o of Figure 2-3. With composition of these mappings as the operation the set $S = \{I,R,R^2,R^3,H,V,P,O\}$ is a group—called the *symmetries of the square.* Notice that this group is noncommutative although all its proper subgroups are commutative. Thus S is our second example of a noncommutative group. The other is the group of Exercise 1-63.

Exercises 2-31 to 2-37

☆*2-31.* Make a table for the group S described above.

2-32. Find all subgroups of S. (There are 10. Use Theorem 2-7.)

2-33. Find the set of all elements that commute with $H \in S$. Find the set of all elements that commute with $R \in S$. Find the set of all elements that commute with every element of S. (These are subgroups by Exercises 2-24 and 2-25.)

2-34. (Symmetries of an equilateral triangle) Consider the rigid motions of an equilateral triangle onto itself. There are six such motions. Make a table for the group of all such motions. Find its subgroups. Note that it is not a commutative group.

2-35. How many rigid motions of a regular pentagon onto itself are there? Of a regular hexagon? Of a regular n-gon?

2-36. Define a one-to-one OP mapping (in the sense of Definition 1-20) of the group of symmetries of the rectangle (Figure 2-2) onto $\langle P(\{a,b\}),\Delta\rangle$ (Exercise 2-27).

2-37. How many symmetries has a cube?

Clock arithmetic groups. The so-called "clock arithmetic" systems are familiar to many students of mathematics at various levels. Some of these systems are groups, and they will be considered next.

Definition 2-6

Consider the set $J_n = \{0, 1, 2, \ldots, n-1\}$ of integers with $n \geq 2$. If $a,b \in J_n$, take the ordinary integer sum (product) of a and b, divide it by n, and consider the least nonnegative remainder (in the sense of Theorem 1-6). This remainder is the **sum modulo** n of a and b (**product modulo** n of a and b) and is denoted by $a +_n b$ $(a \cdot_n b)$.

For example, if $n = 5$, then $J_5 = \{0,1,2,3,4\}$ and

$$2 \cdot_5 3 = 3 +_5 3 = 1$$

To illustrate the "clock arithmetic" nature of this calculation, space marks labeled 0, 1, 2, 3, and 4 equally around a circle; beginning at 0, move three units (in the direction of 1) and then three more units; the terminal mark of this trip will be 1. Similarly, in J_8, $2 +_8 3 = 5$, $7 +_8 6 = 5$, and $5 \cdot_8 6 = 6$.

Theorem 2-9

For every $n \geq 2$ the system $\langle J_n, +_n \rangle$ is an abelian group.

DISCUSSION. If $n = 1$, then $J_1 = \{0\}$, a one-element group, so the theorem actually holds in that case also. The proof of most of the group properties of $+_n$ is actually contained in Exercises 1-84 to 1-86.

PROOF. It follows directly from Definition 2-6 that $+_n$ is commutative in J_n (because $+$ is commutative in J), and from that definition and Theorem 1-6 it follows that $a +_n b \in J_n$ since the least nonnegative remainder r on dividing by n must satisfy $0 \leq r < n$. The identity is 0 since $k + 0 = k +_n 0 = k$ for $k \in J_n$. Also the inverse $-x$ of an element $x \in J_n$ is 0 if $x = 0$ and $n - x$ for all other x's, since $x + (n - x) = n$ and division by n gives the remainder 0.

Since the operation $+_n$ corresponds to the addition operation on equivalence classes defined in Exercise 1-84 for the relation $\equiv_n$ of congruence modulo n on J (with $k \in J_n$ corresponding to E_k), $+_n$ is also associative. ||

Corollary 2-9*a*

There exists a group of order n for every positive integer n.

PROOF. The group $\langle J_n, +_n\rangle$ is an example of a group of order n. ||

We can also form some groups using as the operation multiplication modulo n. To do this, we must first discard the 0 element of J_n—as when forming a multiplicative group of numbers such as $\langle Q \sim \{0\}, \cdot\rangle$. Discarding zero may or may not be sufficient since the system $\langle J_5 \sim \{0\}, \cdot_5\rangle$ is a group but $\langle J_6 \sim \{0\}, \cdot_6\rangle$ is not—since $3 \cdot_6 2 = 0$ and $0 \notin J_6 \sim \{0\}$. To avoid such occurrences we must either delete all those elements of J_n which have a factor in common with n (thus $\langle\{1,5\}, \cdot_6\rangle$ is a group) or use an n such that none of $1, 2, 3, \ldots, n - 1$ has a factor in common with n. Such numbers n are called *primes*.

Definition 2-7

A positive integer p is called a **prime** iff $p \neq 1$ and the equation $n \cdot m = p$ has only the solutions $m = 1$ and $n = p$, or $m = p$ and $n = 1$ in J_+.

As indicated above, certain subsystems of $\langle J_n \sim \{0\}, \cdot_n\rangle$ may be groups. Of course, $\cdot_n$ is always associative in J_n because $\cdot_n$ can be identified with the multiplication of equivalence classes defined in Exercise 1-84.

Exercises 2-38 to 2-45

2-38. Make a table for $\langle J_7 \sim \{0\}, \cdot_7\rangle$. Find the inverse of each element of this group.

2-39. Consider the system $\langle J_n, \cdot_n\rangle$. Show that 1 is the identity. Show also $(n - 1)^{-1} = n - 1$ in J_n.

2-40. Compute $7 +_n 8$, $7 \cdot_n 8$, $3 \cdot_n 3$, $5 +_n 5$, $1 +_n 8$, and $6 \cdot_n 5$ if

a. $n = 9$ b. $n = 12$ c. $n = 15$

2-41. Find all solutions in J_8 of each of the following equations.

a. $2 \cdot_8 x +_8 1 = 4$ b. $x +_8 7 = 4$

c. $4 \cdot_8 x = 0$ d. $x \cdot_8 x = x$

2-42. Find all solutions in J_6 of each of the following equations.

a. $5 \cdot_6 x = 0$ b. $3 \cdot_6 x = 0$

c. $4 +_6 x = 3$ d. $x \cdot_6 x = x$

2-43. Show that $\langle J_{10} \sim \{0\}, \cdot_{10}\rangle$ is not a group but $\langle\{1,3,7,9\}, \cdot_{10}\rangle$ is a group.

2-44. Show that $\langle J_{15} \sim \{0\}, \cdot_{15}\rangle$ is not a group but $\langle\{1,2,4,7,8,11,13,14\}, \cdot_{15}\rangle$ is a group. What is the distinguishing feature of these eight elements of J_{15}?

2-45. Prove that $\langle J_p \sim \{0\}, \cdot_p\rangle$ is a group if p is a prime.

Matrix groups. A person who has already studied matrix algebra will find this next class of examples familiar. He will also know that there are corresponding results for larger matrices than the 2×2 (two-by-two) variety considered here for simplicity.

Definition 2-8

Let $\langle A,+,\ \rangle$ be an algebraic system (with multiplication indicated by juxtaposition). A **2 × 2 matrix** over A is a square array

$$M = \begin{bmatrix} a & b \\ c & d \end{bmatrix}$$

such that $a,b,c,d \in A$. If

$$N = \begin{bmatrix} x & y \\ z & w \end{bmatrix}$$

is also a 2×2 matrix over A, then $M = N$ iff $a = x$, $b = y$, $c = z$, and $d = w$. The **sum** $M + N$ of M and N is defined to be the component-wise sum

$$M + N = \begin{bmatrix} a & b \\ c & d \end{bmatrix} + \begin{bmatrix} x & y \\ z & w \end{bmatrix} = \begin{bmatrix} a + x & b + y \\ c + z & d + w \end{bmatrix}$$

and the **product** MN is defined by

$$MN = \begin{bmatrix} a & b \\ c & d \end{bmatrix}\begin{bmatrix} x & y \\ z & w \end{bmatrix} = \begin{bmatrix} ax + bz & ay + bw \\ cx + dz & cy + dw \end{bmatrix}$$

A device for remembering this somewhat complicated definition of the product is that its first row is formed using the first row of M, turning that row on its side

$$\begin{bmatrix} a \\ b \end{bmatrix}$$

multiplying corresponding elements by the corresponding elements of the jth column ($j = 1,2$) of N, and adding to determine the jth element of the first row of MN; the second row of MN is formed similarly using the second row of M. (This device generalizes nicely to a method for finding products of larger matrices.)

Let $T_2(A)$ denote the set of all 2×2 matrices over A. Our interest here will be in only those situations in which A is a set of numbers.

Definition 2-9

If $M \in T_2(R)$, then the **determinant** of M is the number

$$\det(M) = \det\begin{bmatrix} a & b \\ c & d \end{bmatrix} = ad - bc$$

If $x \in R$, then

$$xM = x\begin{bmatrix} a & b \\ c & d \end{bmatrix} = \begin{bmatrix} xa & xb \\ xc & xd \end{bmatrix}$$

One of the purposes in the discussion of the 2×2 matrices here is to provide an example of a system with a well-behaved addition and a noncommutative multiplication for use in Chapter 4—as well as several examples of groups for the present discussion. In particular, the additive system $\langle T_2(R),+\rangle$ is an abelian group. Since the addition is defined component-wise, associativity and commutativity of addition follow directly from the corresponding properties of R. The identity element is clearly

$$\begin{bmatrix} 0 & 0 \\ 0 & 0 \end{bmatrix}$$

and the inverses (or negatives) are seen to be

$$-\begin{bmatrix} a & b \\ c & d \end{bmatrix} = \begin{bmatrix} -a & -b \\ -c & -d \end{bmatrix}$$

To show that multiplication in $T_2(R)$ is not commutative, let us consider the examples

$$\begin{bmatrix} 1 & 0 \\ 0 & 0 \end{bmatrix}\begin{bmatrix} 0 & 1 \\ 0 & 0 \end{bmatrix} = \begin{bmatrix} 0 & 1 \\ 0 & 0 \end{bmatrix}$$

and

$$\begin{bmatrix} 0 & 1 \\ 0 & 0 \end{bmatrix}\begin{bmatrix} 1 & 0 \\ 0 & 0 \end{bmatrix} = \begin{bmatrix} 0 & 0 \\ 0 & 0 \end{bmatrix}$$

Observe that multiplication in $T_2(R)$ behaves unlike multiplication in R in still another respect: The product of two matrices, neither of which is the zero matrix (additive identity), can be the zero matrix.

Exercises 2-46 to 2-48

2-46. Let

$$A = \begin{bmatrix} 1 & 0 \\ 2 & 1 \end{bmatrix} \qquad B = \begin{bmatrix} 3 & -4\sqrt{2} \\ \sqrt{2} & 1 \end{bmatrix} \qquad C = \begin{bmatrix} \sqrt{2} & -2 \\ \sqrt{3} & -5 \end{bmatrix}$$

Find

a. $A + B$ b. AC c. CA
d. CB e. BC f. $\det(A + B)$
g. $\det(B)$ h. $\det(AC)$ i. $\det(CA)$

2-47. Determine which of the following statements about determinants are true in $T_2(R)$. (Prove or give counterexample.)

a. $\det(M + N) = \det(M) + \det(N)$
b. $\det(MN) = \det(M) \cdot \det(N)$
c. $\det(MN) = \det(NM)$

2-48. Show that the following sets are groups with respect to the multiplication operation in $T_2(R)$.

a. $\{M \mid \det(M) \neq 0\}$ b. $\{M \mid \det(M) = 1\}$

Cartesian product groups. The next example introduces a way of making new groups from old—a way that will be considered at some length in Chapters 6 and 7.

Let $\langle G, \ \rangle$ and $\langle H, \cdot \rangle$ be groups. Then the cartesian product $G \times H$ can be made into a group by using coordinate-wise multiplication; that is,

$$(g_1,h_1) \circ (g_2,h_2) = (g_1g_2,\, h_1 \cdot h_2)$$

Actually, the matrix addition defined for $T_2(R)$ in the preceding example can be thought of as a version of a cartesian product group $R \times R \times R \times R$ using addition as the operation each time. The only difference is that for matrices, the 4-tuples are written

$$\begin{bmatrix} a & b \\ c & d \end{bmatrix}$$

while for the product group in notation similar to that used above, the corresponding element would be (a,b,c,d). In $R \times R \times R \times R$, the group operation would be defined by the equation

$$(a,b,c,d) \circ (x,y,z,w) = (a + x, b + y, c + z, d + w)$$

Exercises 2-49 to 2-50

2-49. Prove that $\langle G \times H, \circ \rangle$, as defined above, is a group.

2-50. When is $\langle G \times H, \circ \rangle$ an abelian group?

Miscellaneous exercises 2-51 to 2-57

2-51. Show that $\{1,-1,i,-i\}$ with ordinary complex number multiplication is a group.

2-52. Let a be the identity in a group $G = \{a,b,c\}$. Show that there is only one way to complete the table for G.

2-53. Let G be a finite group of even order. Show that $x = x^{-1}$ for at least two elements $x \in G$.

☆*2-54.* Let e be the identity in a group $G = \{e,b,c,d\}$ and let $b^{-1} = b$. Show that there are only two ways to complete the table for G and that both yield commutative groups.

2-55. If G is an abelian group then $\{x \mid x = x^{-1}\}$ is a subgroup of G.

2-56. Find the product of all elements of a finite commutative group.

2-57. (For calculus scholars) Let $C[0,1]$ be the set of all continuous real-valued mappings with domain the closed interval $[0,1]$. If $f,g \in C[0,1]$, let $f + g : x \rightarrow xf + xg$. Use theorems of calculus to show that $\langle C[0,1],+\rangle$ is a commutative group and that the set of all differentiable mappings on $[0,1]$ is a subgroup of it.

2-D CYCLIC GROUPS

In the groups $\langle J,+\rangle$ and $\langle J_n,+_n\rangle$ the number 1 plays a particularly significant part in that every element except 0 can be obtained as

$1 + 1 + \cdots + 1$ or $-(1 + 1 + \cdots + 1)$. Thus 1 "generates" the group. To develop this idea in general, we switch from an additive to a multiplicative notation and denote n repetitions of an element x by x^n.

Definition 2-10

Let $\langle G, \ \rangle$ be a group with identity e, and let $x \in G$. Then

1. $x^0 = e$.

2. $x^n = (x^{n-1})x$ for every positive integer n.

3. $x^{-n} = (x^{-1})^n$ for every positive integer n.

(If the operation is denoted by "+," write "nx" instead of "x^n." Note however that $n \notin G$.)

Parts 1 and 2 imply that $x^1 = x^0x = ex = x$.

It would also be helpful to know that part 3 of this definition is consistent with the notation used earlier for inverses; that is, it is necessary to show that the inverse of x is the *minus-one power* of x. To show this, let us temporarily denote the inverse of x by x' and the minus-one power of x by x^{-1}. Then $x^{-1} = (x')^1 = x'$ by (3).

The properties of integral exponents of a group element are those of integral exponents of real numbers.

Theorem 2-10

(1) If $x \in G$ and $n \in J_0$, then $x^{-n} = (x^n)^{-1}$. If $m,n \in J$ and $x \in G$, then (2) $x^mx^n = x^{m+n}$, and (3) $(x^m)^n = x^{mn}$.

PROOF. The first item of business will be to prove by induction that part 2 holds for $m,n \in J_0$.

Let $M = \{n \mid n \in J_0 \text{ and } x^mx^n = x^{m+n} \text{ for all } m \in J_0\}$. Then $x^mx^0 = x^me = x^m = x^{m+0}$, so $0 \in M$. If $r \in M$, that is, if $x^mx^r = x^{m+r}$, then

$$x^mx^{r+1} = x^m(x^rx) = (x^mx^r)x = x^{m+r}x = x^{m+r+1}$$

so that $r + 1 \in M$, too. Thus $M = J_0$, and $x^m x^n = x^{m+n}$ for all $m,n \in J_0$.

We note in passing that positive powers of x commute since

$$x^mx^n = x^{m+n} = x^{n+m} = x^nx^m$$

To prove part 1—again by use of induction—assume that

$$x^{-r} = (x^r)^{-1}$$

(We have already observed this for $r = 1$.) Then

$$\begin{aligned} x^{-(r+1)} &= (x^{-1})^{r+1} && \text{Definition 2-10, part 3} \\ &= (x^{-1})^r x^{-1} && \text{Definition 2-10, part 2} \\ &= (x^r)^{-1} x^{-1} && \text{Induction hypothesis} \\ &= (x x^r)^{-1} && \text{Theorem 2-3} \\ &= (x^{r+1})^{-1} && \text{Definition 2-10, part 2} \end{aligned}$$

Thus part 1 has been proved, and we return to the proof of part 2.

If $-m < 0$ and $-n < 0$ in J, then

$$x^{-m}x^{-n} = (x^{-1})^m(x^{-1})^n = (x^{-1})^{m+n} = x^{-(m+n)} = x^{-m-n}$$

If $m \geq 0$ and $-n < 0$, and $m = n + r$, then

$$x^m x^{-n} = x^{n+r}x^{-n} = x^n x^r x^{-n} = x^r x^n x^{-n} = x^r = x^{m-n}$$

The proof of part 3 is deferred to the exercises. ||

No mention has been made of the rule for exponents of real numbers $(xy)^n = x^n y^n$, because this statement is not true in general. In fact, it is not true for $n = -1$ except in abelian groups, as the earlier study of inverses (particularly Theorem 2-3) showed.

Having established the essential properties of integral exponents in a group, we turn to the study of groups and subgroups which consist of powers of a single element.

Definition 2-11

Let $\langle G, \ \rangle$ be a group and let $g \in G$. Then the **cyclic subgroup generated by** g is

$$[g] = \{g^n \mid n \in J\}$$

If there exists a $b \in G$ such that $[b] = G$, then G is called a **cyclic group.**

To justify the name *cyclic subgroup*, it is necessary to show that $[g]$ is a subgroup of G. This follows largely from Theorem 2-10 and

Theorem 2-6 because $e = g^0 \in [g]$, so $[g] \neq \emptyset$; and if $x,y \in [g]$, then $x = g^m$ and $y = g^n$ for some $m,n \in J$, so

$$xy^{-1} = g^m(g^n)^{-1} = g^m g^{-n} = g^{m-n} \in [g]$$

Thus $[g]$ is a subgroup by Theorem 2-6. Since $g^m g^n = g^n g^m$, $[g]$ is always commutative.

Cyclic groups and cyclic subgroups of other groups may have more than one generator. For example in $\langle J_4,+_4\rangle$, whose table is shown below, $J_4 = [1] = [3]$. Here $[2]$ is the subgroup $\{0,2\}$.

$+_4$	0	1	2	3
0	0	1	2	3
1	1	2	3	0
2	2	3	0	1
3	3	0	1	2

In a finite group G it is clear that the cyclic subgroup $[g]$ is also finite. In an infinite group cyclic subgroups may be finite or infinite. In the group $\langle P(X),\Delta\rangle$ every cyclic subgroup except $[e]$ has order 2 since $A \Delta A = \emptyset$, so that $[A] = \{\emptyset, A\}$. However, in $\langle R_+, \cdot\rangle$ every cyclic subgroup is infinite—with the exception, again, of $[e]$ since e generates only itself.

Definition 2-12

The **order of an element** $g \in G$ is the order of the cyclic subgroup $[g]$ generated by g. If $[g]$ is an infinite set, then g is said to have **infinite order.**

It will be well to be cautious of the word *order* because it is used in two closely related senses. There is the *order of a group* and the *order of an element,* which is, in turn, defined in terms of the order of a subgroup. There is, however, another way of looking at the order of an element which is probably more useful than the definition. The mathematics student encounters it in his first study of the complex numbers when considering the powers of i: $i = i$, $i^2 = -1$, $i^3 = -i$, $i^4 = 1$, $i^5 = i$, $i^6 = -1$, and so on. (See also Exercise 2-51.) The element i has order 4 in the multiplicative group of nonzero complex numbers and $i^4 = 1$, the identity of that group. This is what must happen for elements of finite order.

Theorem 2-11

Let $g \in G$, a group, and let g have finite order n. Then the order of g is the smallest positive integer such that $g^n = e$, the identity.

PROOF. By definition, to say that the order of g is n is to say that $[g]$ contains n elements. What we propose to prove is that these n elements are $g^0 = e, g^1, g^2, \ldots, g^{n-1}$, and we must first show that the n exists.

Consider the $n + 1$ elements $g^0, g^1, \ldots, g^{n-1}, g^n$. Since $[g]$ contains exactly n elements, two of these must be equal. Let $g^r = g^s$ with $0 \leq s < r \leq n$. Then $g^{r-s} = e = g^0$ with $r - s > 0$. Let m be the smallest *positive* integer such that $g^m = e$. If $g^u = g^v$ with $0 \leq u < v < m$, then $g^{v-u} = e$ and $0 < v - u < m$, which contradicts the minimality of m. Thus the elements $g^0, g^1, \ldots, g^{m-1}$ are all distinct.

Consider g^t. Dividing t by m (Theorem 1-6) gives $t = mq + r$ with $0 \leq r < m$. Thus

$$g^t = g^{mq+r} = (g^m)^q g^r = e^q g^r = g^r$$

so that $g^t \in \{g^0, g^1, \ldots, g^{m-1}\}$.

We have thus shown that the elements g^r, $0 \leq r < m$, are all different and every power of g is equal to one of these. Since g generates n distinct elements, $m = n$ and $g^n = e$. ||

Now we have a somewhat simpler way of thinking of the powers of an element of a group. Included in the proof above, at the point at which the division algorithm (Theorem 1-6) was employed, is the assertion that the powers of every group element of finite order "cycle" after the manner of the powers of i. Thus, if g has order 6 in a group, then $g^7 = g$, $g^8 = g^2$, $g^9 = g^3$, and $g^{-1} = g^5$, $g^{-2} = g^4$, and so on.

Notice that the group $\langle J_n, +_n \rangle$ is always a cyclic group of order n. However, not all finite groups are cyclic; not even all finite commutative groups are cyclic. (Example?)

Theorem 2-12

Every subgroup of a cyclic group is cyclic.

This theorem, whose proof is an exercise, says that, if $G = [g]$ and H is a subgroup of G, then $H = [g^r]$ for some $r \in J$.

It follows from the theorems above that every group G has abelian (in fact, cyclic) subgroups, namely, $[g]$ for any $g \in G$. Of course, abelian groups must have only abelian subgroups. We have already considered two non-abelian groups G such that every subgroup, except G itself, is abelian (Tables 1 and 3 of Appendix I).

Further properties of cyclic groups will appear in the next chapter, where it will be shown that, in a sense to be considered there, the groups $\langle J,+\rangle$ and $\langle J_n,+_n\rangle$ are the *only* cyclic groups.

Exercises 2-58 to 2-67

☆*2-58*. Write Definition 2-10 and Theorem 2-10 in additive notation (nx for x^n).

2-59. a. Show that $(ab)^2 = a^2b^2$ for some a and b in the group of Table 3 of Appendix I.

b. For what groups is $(ab)^2 = a^2b^2$ for all a and b? (Prove your assertion.)

2-60. Find the orders of R, H, and P in the group of Table 3 of Appendix I. Find all cyclic subgroups of this group.

2-61. Find all generators of $\langle J_5,+_5\rangle$, of $\langle J_6,+_6\rangle$, and of $\langle\{1,-1,i,-i\},\cdot\rangle$.

2-62. Prove part 3 of Theorem 2-10 (all cases).

2-63. Prove Theorem 2-12. (Let m be the smallest positive integer such that $g^m \in H$.)

☆*2-64*. Let $a,b \in G$, a group. Assume ab has finite order. Show that ba has the same order.

2-65. Find all subgroups of $\langle J_{20},+_{20}\rangle$. (Use Theorem 2-12.)

2-66. In an infinite cyclic group, every subgroup except $\{e\}$ is infinite.

2-67. A *square root* of an element g in a group G is an element $x \in G$ such that $x^2 = g$. Does every element of a group have a square root? Do those which have square roots have exactly two (as in $R \sim \{0\}$)? What general statements about roots are true?

2-E PERMUTATION GROUPS

This section is devoted to the subject of composition of mappings, and to groups whose elements are mappings and whose operation is composition. The mappings considered must have enough in common to make the formation of a group possible. In particular, there must be inverses; but if the inverse of a mapping is a mapping, then the mapping is one-to-one. If the mapping and its inverse have the same domain, then the mapping is *onto* its domain. Thus the principal characteristics of the mappings α in this section are that α is one-to-one and that dm (α) = rn (α).

Theorem 2-13

Let $\mathrm{O}(A)$ be the set of all one-to-one mappings of a set A onto itself. Using composition of mappings as the operation, $\langle \mathrm{O}(A), \quad \rangle$ is a group.

PROOF. That $\langle \mathrm{O}(A), \quad \rangle$ is an algebraic system follows from Exercises 1-32 and 1-33. The operation is associative by Theorem 1-2. The identity mapping $\delta : x \to x$ is in $\mathrm{O}(A)$ and $x(\alpha\delta) = (x\alpha)\delta = x\alpha$, so δ is a right-identity for composition. Finally, if $\alpha \in \mathrm{O}(A)$, then α^{-1} defined by $y\alpha^{-1} = x$ iff $x\alpha = y$ (Definition 1-12) is also in $\mathrm{O}(A)$ and $x(\alpha\alpha^{-1}) = (x\alpha)\alpha^{-1} = y\alpha^{-1} = x$, so $\alpha\alpha^{-1} = \delta$ and α^{-1} is a right-inverse of α. ||

It is groups $\mathrm{O}(A)$ for finite sets A and the subgroups of $\mathrm{O}(A)$ which will be studied here.

Definition 2-13

If A is a finite set, then a one-to-one mapping of A onto itself is called a **permutation** of A. Thus α is a permutation of A iff $\alpha \in \mathrm{O}(A)$ and A is a finite set. Any set of permutations which form a group is called a **permutation group.** The group of all permutations of a set of n elements is called the **symmetric group of degree n** and is denoted by S_n.

In constructing a single permutation of n things, there are n choices for the image of the first thing, $n - 1$ choices for the image of the second thing, etc., so that there are $n(n-1)(n-2) \cdots 2 \cdot 1 = n!$ permutations of a set of n elements. Thus the symmetric group of *degree n* has *order n*!. In discussing these permutations it will usually be most convenient notationally to let the set be $\{1,2,3, \ldots ,n\}$. Thus $S_n = O(\{1,2, \ldots ,n\})$. To develop a permutation notation, begin with the matrix form

$$\alpha = \begin{bmatrix} 1 & 2 & \cdots & n \\ a_1 & a_2 & \cdots & a_n \end{bmatrix}$$

where $a_1,a_2, \ldots ,a_n$ are the numbers $1,2, \ldots ,n$ in some order. For example, in S_4 let

$$\beta = \begin{bmatrix} 1 & 2 & 3 & 4 \\ 2 & 3 & 4 & 1 \end{bmatrix} \quad \text{and} \quad \gamma = \begin{bmatrix} 1 & 2 & 3 & 4 \\ 4 & 2 & 1 & 3 \end{bmatrix}$$

This is interpreted to mean that the image of an element j is written below j. Thus $2\beta = 3$, $2\gamma = 2$, etc. To compute the product $\beta\gamma$, obtain the result of applying $\beta\gamma$ to each element of $\{1,2,3,4\}$. For example, $4(\beta\gamma) = (4\beta)\gamma = 1\gamma = 4$, and

$$\beta\gamma = \begin{bmatrix} 1 & 2 & 3 & 4 \\ 2 & 3 & 4 & 1 \end{bmatrix}\begin{bmatrix} 1 & 2 & 3 & 4 \\ 4 & 2 & 1 & 3 \end{bmatrix} = \begin{bmatrix} 1 & 2 & 3 & 4 \\ 2 & 1 & 3 & 4 \end{bmatrix}$$

However,

$$\gamma\beta = \begin{bmatrix} 1 & 2 & 3 & 4 \\ 1 & 3 & 2 & 4 \end{bmatrix} \neq \beta\gamma$$

A little reflection should lead to methods of reading the product $\beta\gamma$ from the matrices for β and γ without any intermediate writing.

This notation can be simplified by first ignoring elements that are mapped onto themselves and then using the fact that if $a_1\alpha = a_2$, $a_2\alpha = a_3$, and so forth, the element a_1 will eventually show up again as an image of some a_j ($a_j\alpha = a_1$) because A is a finite set. Thus the *cycle notation* $\alpha = (a_1a_2 \cdots a_j)$ can be used to denote this when interpreted as $a_1\alpha = a_2$, $a_2\alpha = a_3$, $\ldots$, $a_j\alpha = a_1$. (See Figure 2-4.) The cycle notation for β and γ above would be $\beta = (1234)$ and $\gamma = (143)$. If a permutation turns out not to be representable as a single cycle, it can be represented as a product of *disjoint cycles* (cycles with no common elements). To see how this is done, consider the permutation

$$\alpha = \begin{bmatrix} 1 & 2 & 3 & 4 & 5 & 6 & 7 & 8 \\ 2 & 1 & 3 & 7 & 4 & 8 & 5 & 6 \end{bmatrix}$$

Consider first the element 1. Since $1\alpha = 2$ and $2\alpha = 1$, the cycle (12) gives this information. Because $3\alpha = 3$, we skip 3 and move on to 4. The cycle (475) describes the fact that $4\alpha = 7$, $7\alpha = 5$, and $5\alpha = 4$. Finally "$6\alpha = 8$ and $8\alpha = 6$" becomes "(68)" in cycle notation. Thus $\alpha = (12)(475)(68)$ as a product of disjoint cycles. It should be clear that disjoint cycles commute, so $\alpha = (475)(68)(12)$ also. Note, in addition, that the representation of a single cycle is not unique;

(13427) means

(142) (35) means

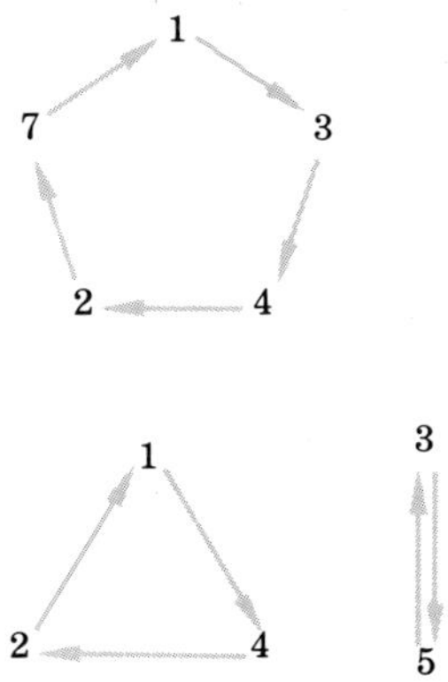

Figure 2-4 Cycles.

it is necessary to preserve only the cyclic order of the elements. That is, $(475) = (754) = (547)$. One could also write $\alpha = (12)(86)(754)$.

If a permutation happens to be the identity

$$\delta = \begin{bmatrix} 1 & 2 & \cdots & n \\ 1 & 2 & \cdots & n \end{bmatrix}$$

it would be a little awkward to skip all the elements in forming a cycle, so it is customary to write "(1)" as the cycle form of the identity permutation. Clearly "(3)" or "(n)" would do as well.

Definition 2-14

A cycle $(a_1a_2 \cdots a_k)$ is called a **cycle of length k** iff it contains k elements. A cycle of length 2 is called a **transposition.** Every cycle of length 1 denotes the identity permutation.

The procedure used above to write the permutation

$$\alpha = (12)(475)(68)$$

as a product of disjoint cycles can be generalized to prove that every permutation can be expressed as a product of disjoint cycles. The key to the proof is the fact that a permutation is a mapping of a finite set, so the elements eventually are exhausted. Also the facts that a permutation is one-to-one and onto imply that the cycle $(a_1a_2 \cdots a_k)$ must close with some a_k mapping to a_1 and cannot close with a_k mapping to some intermediate element a_i $(i = 2,3, \ldots ,k)$.

Theorem 2-14

Every cycle can be written as a product of transpositions.

PROOF. This can be done to the cycle $(a_1a_2 \cdots a_n)$ merely by letting the first element in each transposition be a_1 and the second be the elements $a_2,a_3, \ldots ,a_n$ in order. Thus

$$(a_1a_2a_3 \cdots a_{n-1}a_n) = (a_1a_2)(a_1a_3) \cdots (a_1a_{n-1})(a_1a_n) \quad \|$$

So (1432) = (14)(13)(12), but this is not a unique representation even with respect to the number of transpositions since

$$(1432) = (14)(24)(13)(24)(12)$$

(Verifying these calculations is one of the best ways to get practice in computing products of cycles. Transpositions are about the trickiest of cycles.) A bit of advice that seems to overcome many errors made in computing products of cycles is "keep moving." Thus in computing (142)(345)(12), remember that once it has been discovered that 1 maps to 4, the image of 4 comes from the *next* cycle or a following one, not from the same cycle. Thus $1 \to 4$, and $4 \to 5$, and 5 is left fixed by the third factor, so $1 \to 5$.

Since a transposition is designed to transpose (interchange) two symbols and leave all others fixed, the following information can be inferred from Theorem 2-14: An ordering of a finite set can be obtained from any other ordering by successive interchanges of the positions of two elements. The factoring of a permutation as a product of transpositions may not be unique, as was observed earlier, but there is something unique about the factoring. The number of transpositions is always even or always odd.

Definition 2-15

A cycle (or permutation) is called **even** iff it can be written as a product of an even number of transpositions. A cycle (or permutation) is **odd** iff it can be written as a product of an odd number of transpositions.

Theorem 2-14 says, in part, that every cycle is either even or odd. The next theorem says that it cannot be both.

Theorem 2-15

No cycle is both even and odd.

DISCUSSION. The proof of this theorem is a classic example of an indirect attack on a problem. What we shall do can best be illustrated by an example. Consider the product

$$(1 - 2)(1 - 3)(1 - 4)(2 - 3)(2 - 4)(3 - 4) = 12$$

of all factors $(a - b)$ with $a,b \in \{1,2,3,4\}$ and $a < b$. Interchange 1 and 3 (that is, replace every occurrence of 1 by 3 and vice versa). This yields

$$(3 - 2)(3 - 1)(3 - 4)(2 - 1)(2 - 4)(1 - 4) = -12$$

The reader should satisfy himself that the other transpositions (12), (23), (14), (24), (34) of $\{1,2,3,4\}$ also change the original product to -12.

PROOF. Consider the permutation $\alpha = (a_1 \cdots a_n)$ where the a_i's are distinct positive integers. Write α as products $\beta_1\beta_2 \cdots \beta_k$ and $\gamma_1\gamma_2 \cdots \gamma_m$ of transpositions. Now form the product

$$p = \prod_{i<j} (a_j - a_i)$$

the product of all the differences of the numbers a_i and a_j in which $i < j$. Since the a_i's are distinct, $p \neq 0$. Now perform the transposition $\beta_1 = (rs)$, with $r < s$, in p. This reverses the sign of the factor $(a_r - a_s)$. If neither a_r nor a_s appears in a factor $(a_i - a_j)$, that factor is unchanged. We can pair together the products of factors $(a_r - a_t)(a_t - a_s)$ if $r < t < s$ (and the other possibilities for $t \neq r$ and $t \neq s$) and observe that the product is unchanged when a_r and a_s are interchanged. Thus β_1, and every transposition, changes p to $-p$. Thus $\beta_1\beta_2 \cdots \beta_k$ changes p to $(-1)^k p$ and $\gamma_1\gamma_2 \cdots \gamma_m$ changes p to $(-1)^m p$. But these are both the same permutation α and must have the same total effect on p. Thus $(-1)^k p = (-1)^m p$, and k and m are either both even or both odd. ||

Corollary 2-15*a*

No permutation is both even and odd.

This result can be used to show that certain configurations of things are impossible in various games or puzzles such as those puzzles which involve rearranging numbered figures within specified boundaries.

Before leaving the discussion of cycles, let us look at one symmetric group S_3. The $3! = 6$ permutations of $\{1,2,3\}$ are:

$$(1) = \begin{bmatrix} 1 & 2 & 3 \\ 1 & 2 & 3 \end{bmatrix} \qquad (12) = \begin{bmatrix} 1 & 2 & 3 \\ 2 & 1 & 3 \end{bmatrix} \qquad (13) = \begin{bmatrix} 1 & 2 & 3 \\ 3 & 2 & 1 \end{bmatrix}$$

$$(23) = \begin{bmatrix} 1 & 2 & 3 \\ 1 & 3 & 2 \end{bmatrix} \qquad (123) = \begin{bmatrix} 1 & 2 & 3 \\ 2 & 3 & 1 \end{bmatrix} \qquad (132) = \begin{bmatrix} 1 & 2 & 3 \\ 3 & 1 & 2 \end{bmatrix}$$

The table for S_3 is given below.

	(1)	(12)	(13)	(23)	(123)	(132)
(1)	(1)	(12)	(13)	(23)	(123)	(132)
(12)	(12)	(1)	(123)	(132)	(13)	(23)
(13)	(13)	(132)	(1)	(123)	(23)	(12)
(23)	(23)	(123)	(132)	(1)	(12)	(13)
(123)	(123)	(23)	(12)	(13)	(132)	(1)
(132)	(132)	(13)	(23)	(12)	(1)	(123)

Note that the even permutations (1), (123), and (132) form a subgroup of S_3. There are also three subgroups of order 2 and all proper subgroups are cyclic. If this begins to sound like the same sort of properties we discovered earlier in the group of Table 1 of Appendix I, then the time is ripe to move on to the next chapter and discuss similarities between groups in more detail.

Exercises 2-68 to 2-80

2-68. Write the following permutations as products of disjoint cycles.

a. $\begin{bmatrix} 1 & 2 & 3 & 4 & 5 & 6 \\ 4 & 1 & 3 & 6 & 2 & 5 \end{bmatrix}$ b. $\begin{bmatrix} 1 & 2 & 3 & 4 & 5 & 6 \\ 4 & 1 & 5 & 2 & 6 & 3 \end{bmatrix}$

☆c. $\begin{bmatrix} 1 & 2 & 3 & 4 & 5 & 6 & 7 & 8 \\ 8 & 5 & 7 & 4 & 1 & 6 & 3 & 2 \end{bmatrix}$ d. $\begin{bmatrix} 1 & 2 & 3 & 4 & 5 & 6 & 7 & 8 \\ 4 & 8 & 3 & 1 & 5 & 6 & 7 & 2 \end{bmatrix}$

2-69. Compute the products (composites) of the cycles. Write each as a product of disjoint cycles or as a single cycle.

a. (12)(13)(452) b. (1452)(43125)
☆c. (254)(143)(21) d. (14356)(12356)
e. (12)(13)(14)(15) f. (146)(1)(35)(3)

2-70. Write the following permutations in matrix form as elements of S_5.

a. (1342) b. (24)(315) c. (312)
d. (1) e. (12453)

2-71. Write each of the following permutations as the product of two different numbers of transpositions.

a. (1432)(65) b. (142)(735)

2-72. Prove the corollary to Theorem 2-15.

2-73. Show that every transposition in S_n has order 2. Show that every cycle of length k has order k.

2-74. Show that the inverse of the cycle $(a_1a_2 \cdots a_n)$ is the cycle $(a_na_{n-1} \cdots a_2a_1)$.

2-75. Find $(1234)^2$, $(1234)^3$, and $(1234)^{-1}$. Are any of these equal?

☆*2-76.* Show that exactly half of the $n!$ elements of S_n are even permutations.

2-77. Let A_n be the set of all even permutations in S_n. Show that A_n

is a subgroup of S_n. (A_n is called the *alternating group* of degree n.)

2-78. List the elements of the cyclic subgroup $[(124)]$ of S_4.

2-79. Number the vertices of a square by 1, 2, 3, 4 beginning at the upper right and moving counterclockwise. Show that each element of the group of symmetries of the square can be considered as a permutation of the vertices, and exhibit the subgroup of S_4 thus obtained.

2-80. A group is the *dihedral group* D_n iff there exist elements $x,y \in D_n$ such that $xy \neq yx$, $x^2 = e$, $y^n = e$, and every element of D_n can be expressed as a product $x^r y^s$ with $0 \leq r < 2$ and $0 \leq s < n$.

☆a. Show that these defining conditions are inconsistent for $n = 1$ and 2.

b. How many elements does D_n have ($n \geq 3$)?

c. Exhibit a permutation group, a subgroup of S_n, which satisfies the definition of D_n. (Note that the group of symmetries of the square satisfies the definition of D_4.)

3 SUBGROUPS AND HOMOMORPHISMS

It is the purpose of this chapter to gather together in a more intelligible package several ideas that have been considered already. One of these ideas is the similarity of behavior of some of the groups discussed earlier; another is the idea of operation-preserving mappings between groups (discussed in Chapter 1); a third is the relationship between a group and its subgroups.

If we consider the two groups $\langle J_4, +_4\rangle$ and $\langle J_5 \sim \{0\}, \cdot_5\rangle$, both with four elements, their tables look like

$+_4$	0	1	2	3
0	0	1	2	3
1	1	2	3	0
2	2	3	0	1
3	3	0	1	2

and

$\cdot_5$	1	2	3	4
1	1	2	3	4
2	2	4	1	3
3	3	1	4	2
4	4	3	2	1

If we now interchange the headings 3 and 4 in the second table to get

$\cdot_5$	1	2	4	3
1	1	2	4	3
2	2	4	3	1
4	4	3	1	2
3	3	1	2	4

the similarity is very striking. In fact, if we were to replace each occurrence of 1, 2, 4, and 3 in this table by 0, 1, 2, and 3, respectively, we would have precisely the first table. In other words, the structure of these two groups is the same. They differ only in the labels given to the elements and operations.

If we consider $\langle J_2, +_2\rangle$ and $\langle J_4, +_4\rangle$, they cannot possibly be identical because they have different numbers of elements. But a replacement of 0 and 2 by 0 and of 1 and 3 by 1 in the table for J_4 has the following effect:

$+_4$	0	1	2	3
0	0	1	2	3
1	1	2	3	0
2	2	3	0	1
3	3	0	1	2

becomes

$+_2$	0	1	0	1
0	0	1	0	1
1	1	0	1	0
0	0	1	0	1
1	1	0	1	0

which gives exactly the same information as

$+_2$	0	1
0	0	1
1	1	0

Putting the matter another way, if the elements of J_4 are put in the order 0, 2, 1, 3, the table for J_4 becomes

$+_4$	0	2	1	3
0	0	2	1	3
2	2	0	3	1
1	1	3	2	0
3	3	1	0	2

which, in 2 × 2 blocks, can be identified as above with the table for J_2. So there is a structural relationship between these groups of a kind that identifies *subsets* of J_4 with *elements* of J_2. (Observe that $\{0,2\}$, which is identified with $0 \in J_2$, is a subgroup of J_4.)

These two examples are examples of the concepts called *isomorphism* and *homomorphism*, respectively. In this chapter the theoretical aspects of these ideas will be developed. We shall see also that the partitioning of group tables into blocks as above cannot be accomplished for every subgroup of every finite group but that the subgroups for which this is possible are particularly significant.

3-A HOMOMORPHISMS

A homomorphism is just an operation-preserving mapping as defined in Chapter 1.

Definition 3-1

Let $\langle G,\circ\rangle$ be a group and $\langle A, \cdot \rangle$ an algebraic system. A mapping $\alpha : G \rightarrow A$ is called a **homomorphism** of G to A iff

$$(a \circ b)\alpha = a\alpha \cdot b\alpha \text{ for all } a,b \in G$$

Theorem 3-1

If α is a homomorphism of the group $\langle G,\circ\rangle$ to the system $\langle A, \cdot \rangle$, then the subsystem $\langle \text{rn } (\alpha), \cdot \rangle$ in A is a group. (The homomorphic image of a group is a group.)

PROOF. This "proof" merely collects facts from Theorems 1-3 and 1-5. By Theorem 1-3, $\langle \text{rn } (\alpha), \cdot \rangle$ is an algebraic system. By Theorem 1-5 (Exercises 1-68, 1-69, and 1-70) it is associative, has an identity, and has inverses for each element because G has the corresponding properties. ||

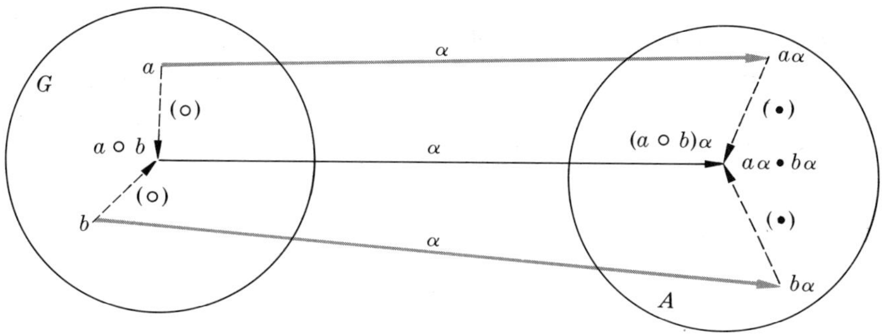

Figure 3-1 ***Homomorphism.***

Theorem 1-4 implies that $\langle \text{rn } (\alpha), \cdot \rangle$ will be an abelian group if $\langle G,\circ\rangle$ is abelian. Exercise 1-70 has the added force of stating that $a^{-1}\alpha = (a\alpha)^{-1}$ in A for all $a \in G$. It is important to recognize that the defining equation, $(a \circ b)\alpha = a\alpha \cdot b\alpha$, is an equation in the system A. The equation says that if one forms the product of any two elements of G and takes the image of that product in A, the result is the same as if it had been obtained by finding the images in A of the two elements and taking the product in A of those images. So if a mapping is a homomorphism, then "multiply and map" is the same as "map and multiply." (See Figure 3-1.)

Theorem 3-1 makes it possible to simplify the complexity of the image set slightly. It is possible without any loss of generality to discuss "a homomorphism of a group *onto a group*" instead of "a homomorphism of a group to an algebraic system." If there is a homomorphism of a group G onto a group $\bar{G}$, we shall say that G and $\bar{G}$ are *homomorphic.* The relation of being homomorphic is reflexive and transitive but not symmetric in a collection of groups since the image $\bar{G}$ may be a smaller set than G—as in the second example at the beginning of the chapter.

Definition 3-2

A one-to-one homomorphism α of a group G to a group $\bar{G}$ is called an **isomorphism,** and the groups G and $rn(\alpha)$ are said to be **isomorphic.**

Isomorphic groups are "algebraically identical" because they differ only in labeling. The groups $\langle J_4, +_4\rangle$ and $\langle J_5 \sim \{0\}, \cdot_5\rangle$ with which this chapter began are isomorphic. The relation of being isomorphic is an equivalence relation on any set of groups.

Exercises 3-1 to 3-13

3-1. Do any of the following exercises not done earlier.
a. 1-68 b. 1-69 c. 1-70

3-2. Show that "homomorphic" is a reflexive and transitive relation on any nonempty set of groups.

3-3. Show that "isomorphic" is an equivalence relation on any nonempty set of groups.

3-4. Let $x\alpha = \log_{10} x$. Show that α is an isomorphism of $\langle R_+, \cdot\rangle$ onto $\langle R, +\rangle$. (Show one-to-one and onto graphically.)

3-5. Show that the mapping β of $\langle J, +\rangle$ onto $\langle J_4, +_4\rangle$ defined by $n\beta = r$, where r is the least nonnegative remainder on dividing n by 4, is a homomorphism.

3-6. Is $\langle J_4, +_4\rangle$ isomorphic to $\langle\{1, -1, i, -i\}, \cdot\rangle$? Is $\langle J_4, +_4\rangle$ isomorphic to the group $\langle \text{P}(\{a,b\}), \Delta\rangle$ of Exercise 2-27? Is $\langle \text{P}(\{a,b\}), \Delta\rangle$ isomorphic to the group of symmetries of a rectangle?

☆***3-7.*** Show that every cyclic group of order n is isomorphic to $\langle J_n, +_n \rangle$. *Corollary.* Two finite cyclic groups of the same order are isomorphic. (Proof by Exercises 3-7 and 3-3.)

3-8. Show that any two infinite cyclic groups are isomorphic by showing them isomorphic to $\langle J, + \rangle$.

3-9. Define a homomorphism from $\langle J_6, +_6 \rangle$ onto:
a. $\langle J_3, +_3 \rangle$ b. $\langle J_2, +_2 \rangle$

3-10. Define a homomorphism from $\langle J_{2n}, +_{2n} \rangle$ onto $\langle J_n, +_n \rangle$.

3-11. Are $\langle J_3, +_3 \rangle$ and $\langle J_2, +_2 \rangle$ homomorphic?

3-12. Consider four groups of order 8: (1) The symmetries of the square, (2) $\langle P(\{1,2,3\}), \Delta \rangle$, (3) $\langle J_8, +_8 \rangle$, and (4) $\langle \{1,2,4,7,8,11,13,14\}, \cdot_{15} \rangle$. Show that no two of these are isomorphic. (See the tables in Appendix I.)

☆*3-13.* Show that S_3 (the symmetric group of degree 3) and the group $\langle F, \ \rangle$ of Table 1 of Appendix I are isomorphic.

▲

The study of the theory of groups developed historically as the study of one-to-one mappings of a set A onto itself (called *permutations* if A is finite). Actually this theory of groups of mappings (as in Theorem 2-13) and the abstract theory of groups are effectively identical. This was proved by one of the pioneers of abstract algebra, the British mathematician Arthur Cayley (1821–1895).

Theorem 3-2 (Cayley)

Every group is isomorphic to a group of one-to-one mappings. (Every finite group is isomorphic to a permutation group.)

Exercises 3-14 to 3-20

3-14 to 3-16. The following exercises and commentary constitute a proof of Cayley's theorem.

3-14. Let $\langle G, \circ \rangle$ be a group. The first need is for a set on which to define a group of one-to-one onto mappings. The set G is the only one available, so we let τ_a be defined for each $a \in G$ by:

$x\tau_a = x \circ a$ for all $x \in G$. Show that τ_a is a one-to-one mapping of G onto itself.

☆*3-15.* Let $T = \{\tau_a \mid a \in G\}$. Define the mapping $\alpha : G \to T$ by $a\alpha = \tau_a$ for all $a \in G$. Show that α is a one-to-one mapping of G onto T.

3-16. Show that $(a \circ b)\alpha = (a\alpha)(b\alpha)$; that is, $\tau_{a \circ b} = \tau_a \tau_b$ in $\langle T, \ \rangle$. ||

3-17. Find a permutation group isomorphic to $\langle J_3, +_3\rangle$. (Use permutations of $\{0,1,2\}$ for convenience.)

3-18. Find a permutation group isomorphic to the group of symmetries of a rectangle.

3-19. Find a permutation group isomorphic to $\langle J_4, +_4\rangle$.

3-20. Why does not the equation $x\tau_a = a \circ x$ define an isomorphism of G onto T?

3-B COSETS

To study subgroups at this point may appear to be quite a change from studying homomorphisms, but the goal of the chapter is to synthesize the two concepts and, in fact, to establish a one-to-one correspondence between the homomorphisms of a group and a class of subgroups of the group—the "normal" subgroups.

Definition 3-3

Let $\langle G, \circ\rangle$ be a group, A a subset of G, and $x \in G$. Then the sets $A \circ x$ and $x \circ A$ are defined by

$$A \circ x = \{a \circ x \mid a \in A\} \qquad \text{and} \qquad x \circ A = \{x \circ a \mid a \in A\}$$

If H is a *subgroup* of G, then $x \circ H$ is called the **left coset** of x and H, and $H \circ x$ is the **right coset** of x and H, in G.

To form a left coset, take a subgroup $H \subset G$ and an element $x \in G$ and form the set of all products $x \circ h$ such that $h \in H$. (If the operation is denoted by $+$, the left coset will be denoted by $x + H$; if the operation is denoted by juxtaposition, the left coset will be denoted

xH, etc.) For example, if H is the subgroup $H = \{a,b\}$ of the group F of Table 1 of Appendix I (which is reproduced below) then $cH = \{ca,cb\} = \{c,d\}$, $eH = \{ea,eb\} = \{e,f\}$, and $He = \{ae,be\} = \{e,c\}$. Since $eH \neq He$, the left and right cosets determined by the same element and subgroup may be different sets. This cannot happen, of course, if the group is commutative. In the same example,

$$bH = \{ba,bb\} = \{b,a\} = H$$

and it is true in general that, if $b \in H$, then $bH = Hb = H$. Here are two examples of left cosets of infinite subgroups: (1) The set $H = \{0, \pm 5, \pm 10, \ldots\}$ is a subgroup of $\langle J, +\rangle$ and

$$2 + H = H + 2 = \{\ldots, -8, -3, 2, 7, 12, \ldots\}$$

the set of all integers which are 2 greater than a multiple of 5. (2) The set Q is a subgroup of $\langle R,+\rangle$ and $\sqrt{3} + Q$ is the set of all real numbers obtained by adding $\sqrt{3}$ to a rational number.

To analyze what may or may not happen when considering cosets, consider the list below of all the left and right cosets of the subgroups $H = \{a,b\}$ and $K = \{a,d,e\}$ of the group $\langle F, \ \rangle$.

F:	a	b	c	d	e	f
a	a	b	c	d	e	f
b	b	a	e	f	c	d
c	c	d	a	b	f	e
d	d	c	f	e	a	b
e	e	f	b	a	d	c
f	f	e	d	c	b	a

$H = \{a,b\}$
$aH = \{a,b\} = Ha = \{a,b\}$
$bH = \{b,a\} = Hb = \{b,a\}$
$cH = \{c,d\} \neq Hc = \{c,e\}$
$dH = \{d,c\} \neq Hd = \{d,f\}$
$eH = \{e,f\} \neq He = \{e,c\}$
$fH = \{f,e\} \neq Hf = \{f,d\}$

$K = \{a,d,e\}$
$aK = \{a,d,e\} = Ka = \{a,d,e\}$
$bK = \{b,f,c\} = Kb = \{b,c,f\}$
$cK = \{c,b,f\} = Kc = \{c,f,b\}$
$dK = \{d,e,a\} = Kd = \{d,e,a\}$
$eK = \{e,a,d\} = Ke = \{e,a,d\}$
$fK = \{f,c,b\} = Kf = \{f,b,c\}$

Several important things happen here:

1. Among these, the collection of all left cosets (or of all right cosets) of a given subgroup is a partition of the group. In the case of the left cosets of H, $aH = bH$, $cH = dH$, and $eH = fH$; but aH, cH, and eH are disjoint; and $aH \cup cH \cup eH = F$.
2. Also true are that $xH = yH$ iff $x \in yH$, and that $xK = yK$ iff $x \in yK$ for all $x,y \in F$.
3. Something which is true for K but not for H is that every left coset is equal to the right coset determined by the same element. In this case, if one selects an element from aK and one from bK and multiplies them, the product is in bK; if one selects both elements from aK or both from bK, the product is in aK; so the set containing the product depends only on the sets from which the elements come, not on the elements themselves. (This is clearer if the headings in the table for $\langle F,\ \rangle$ are in the order a, d, e, b, c, f. Note the last form of the table for $\langle J_4, +_4\rangle$ at the beginning of this chapter.) This is not true for the partition of F by left cosets of H.

We shall soon show that things like these happen generally and develop the theory largely around the better behaved subgroups like K.

Exercises 3-21 to 3-24

3-21. Write out the left cosets of the subgroup $\{I,R^2\}$ of Table 3 of Appendix I.

3-22. For the subgroup $\{I,H\}$ of the group of Table 3 of Appendix I, find an x such that $x\{I,H\} \neq \{I,H\}x$.

3-23. Describe the left cosets of $H = \{0, \pm 5, \pm 10, \ldots\}$ in $\langle J,+\rangle$. If $a + H$ and $b + H$ are left cosets of H, define a "sum" of cosets by $(a + H) + (b + H) = (a + b) + H$. Make a table for $+$. Is this system isomorphic to any previously studied system?

3-24. List the left cosets of the subgroup $H = \{0,5,10,15\}$ in $\langle J_{20}, +_{20}\rangle$. Do the same for $K = \{0,4,8,12,16\}$. Make tables for the "sum" of cosets, as defined in Exercise 2-23, in each case. How do these compare with tables for H and K?

▲

The next theorem is a basic tool for the manipulation and identification of cosets. The information accumulated here will be valuable in establishing most of the results relating subgroups to homomorphisms.

Theorem 3-3

Let H be a subgroup of a group $\langle G, \ \rangle$ and let $a,b \in G$. Then the following three conditions are equivalent: $aH = bH$, $a^{-1}b \in H$, $a \in bH$. In addition, the relation ρ defined by

$$a \,\rho\, b \text{ iff } aH = bH$$

is an equivalence relation in G and the equivalence classes E_a determined by ρ are the left cosets of H in G. ($E_a = aH$.)

PROOF. To show the three conditions equivalent, we shall show that $aH = bH$ implies $a^{-1}b \in H$ implies $a \in bH$ implies $aH = bH$.

If $aH = bH$, since $e \in H$, there exists $h \in H$ such that $ah = be$. Multiplying on the left by a^{-1}, $h = a^{-1}be = a^{-1}b$. Thus $a^{-1}b \in H$. If $a^{-1}b = h$ with $h \in H$, then $b = ah$ and $bh^{-1} = a$. Since H is a subgroup, $h^{-1} \in H$. Thus $bh^{-1} \in bH$ and $a \in bH$.

If $a \in bH$, then there exists $h \in H$ such that $a = bh$. If $x \in aH$, then there exists $k \in H$ such that $x = ak$. Thus

$$x = ak = (bh)k = b(hk) \in bH$$

since H is closed under the group operation. We have proved $aH \subset bH$. If $a = bh$ and $y \in bH$, then there exists $k \in H$ such that $y = bk$. Then $y = bk = (ah^{-1})k = a(h^{-1}k) \in aH$. Thus we have also proved $bH \subset aH$, so that $aH = bH$.

The relation ρ on G defined by $aH = bH$ is an equivalence relation because equality is an equivalence relation. If $a \,\rho\, b$, then $aH = bH$, so that $a \in bH$ by the first part of the theorem—and conversely. Thus $a \,\rho\, b$ iff $a \in bH$. Therefore, $E_b = bH$. ||

This theorem yields quite a lot of corollary information. First, since the equivalence classes of an equivalence relation form a partition, the left cosets of H in G determined by elements a and b are either identical or disjoint. Also a left coset of H may be called xH if, and only if, x is an element of xH. Second, although the left cosets aH are sets and H is a *set* in the symbol aH, H behaves much like a single element. In particular, $aH = bH$ iff $H = a^{-1}bH$. (To prove this, let $aH = bH$. Then $a^{-1}b \in H$. But $a^{-1}b \in H$ iff $a^{-1}bH$ is a name for the coset H.) In this connection, if $H^{-1} = \{h^{-1} \mid h \in H\}$, then $H^{-1} = H$ since H is a subgroup, and $HH^{-1} = HH = H$.

The left cosets of a subgroup H of group G are a rather special kind of partition since all left cosets of H are the same size (in the sense of existence of one-to-one mappings from one to another).

Theorem 3-4

Let H be a subgroup of a group $\langle G, \ \rangle$, and let $a \in G$. Then the mapping $\alpha : h \to ah$ is a one-to-one mapping from H onto the coset aH.

PROOF. (Onto) If $y \in aH$, then $y = ah$ for some $h \in H$ by the definition of aH. Thus for that $h \in H$, $h\alpha = ah = y$.

(One-to-one) If $h\alpha = h'\alpha$ in aH, then $ah = ah'$, so that $h = h'$ by left cancellation. ||

Corollary 3-4*a*

There is a one-to-one mapping from any left coset of H onto any other left coset of H.

PROOF. The mapping $\alpha : h \to ah$ is a one-to-one mapping from H onto aH; $\beta : h \to bh$ is a one-to-one mapping from H onto bH. Thus $\alpha^{-1}\beta$ is a one-to-one mapping from aH onto bH—taking ah to bh. ||

This result leads to a rather interesting property of subgroups of finite groups—a property that may have been observed earlier when all of the subgroups of groups of order 8 had order 1, 2, 4, or 8, and all the subgroups of groups of order 6 had order 1, 2, 3, or 6.

Definition 3-4

The number of left cosets of a subgroup H of a group G is called the **index** of H in G.

Theorem 3-5 (Lagrange)

The order of a subgroup H of a finite group G divides the order of G. (More precisely, if n is the order of G, m the order of H, and q the index of H in G, then $n = mq$.)

PROOF. Let $a_1H, a_2H, \ldots, a_qH$ be the distinct left cosets of H in G. (One of these is, of course, H itself.) By Theorem 3-4 each coset contains exactly m elements (the order of H). By Theorem 3-3 the collection $\{a_iH \mid i = 1,2, \ldots, q\}$ is a partition of G, so

$$G = a_1H \cup a_2H \cup \cdots \cup a_qH$$

has been divided into q disjoint subsets each of which contains m elements. Thus there are $m \cdot q$ elements in G; so $m \cdot q = n$, and m divides n. ||

Lagrange's theorem provides some additional information about cyclic groups and cyclic subgroups.

Corollary 3-5*a*

Every finite group of prime order is cyclic and every element except the identity is a generator.

Corollary 3-5*b*

If G is a group of order n, then $a^n = e$, the identity, for every $a \in G$.

The proofs of these corollaries are deferred to the exercises. Corollary 3-5*a* is particularly interesting in combination with Exercise 3-7, which says that two finite cyclic groups of the same order are isomorphic. It follows, since every group of prime order is cyclic, that there is exactly one group of a given prime order—up to isomorphism. That is, any two groups of the same prime order p are (1) cyclic and (2) isomorphic, so that $\langle J_p, +_p\rangle$ is effectively the only group of order p. Thus for the smaller groups, there is effectively only one group for each of the orders 1, 2, 3, 5, and 7. There are two nonisomorphic groups of order 4 (Exercise 2-54), and both of these are abelian. Since every cyclic group is abelian, the smallest non-abelian group must have order at least 6. Thus S_3 is the smallest non-abelian group. (We have not answered the question of the existence of a non-abelian group of order 6 which is not isomorphic to S_3.)

Exercises 3-25 to 3-31

3-25. Write Theorem 3-3 for right cosets.

☆*3-26.* Prove Corollary 3-5*a*.

3-27. Prove Corollary 3-5*b*.

3-28. Show that A_4, a group of order 12 (see Exercise 2-77), has no subgroup of order 6. Thus a finite group may not have a subgroup of every order which divides the order of the group.

3-29. If G is a cyclic group of order n, and m divides n, then G has a subgroup of order m.

3-30. Find a subgroup of every order d such that d divides 8 in the group $\langle\{1,2,4,7,8,11,13,14\}, \cdot_{15}\rangle$.

3-31. Let H be a subgroup of index 2 in a group G. Show that $aH = Ha$ for every $a \in G$.

3-C NORMAL SUBGROUPS

An example involving subgroups H and K of the group $F = \{a,b,c,d,e,f\}$ was written out in detail in the last section. It was the case there that $xK = Kx$ for all $x \in F$ but that $yH \neq Hy$ for some $y \in F$. This section and the next study properties of those subgroups of a group which behave as K does.

Definition 3-5

Let H be a subgroup of a group $\langle G,\circ\rangle$. Then H is **normal** in G iff $a \circ H = H \circ a$ for all $a \in G$.

Exercise 3-31 declares that every subgroup of index 2 is normal. Every group G has at least two normal subgroups, the improper subgroups G and $\{e\}$. In Exercise 2-25 it was shown that the subset $C = \{x \mid xa = ax \text{ for all } a \in G\}$ is a subgroup of a group G. This subgroup is clearly normal in any group, and every subgroup of an abelian group is normal. However, the statement that $xH = Hx$ does not necessarily mean that $xh = hx$ for every $h \in H$. The equality is a set equality and means just that the elements of xH are the same as those of Hx. In other words $xH = Hx$ iff for every $h \in H$ there exists $k \in H$ such that $xh = kx$, and vice versa.

A concept equivalent to that of normality is *invariance* of subgroups. We approach this concept through the idea of automorphisms.

Definition 3-6

An **automorphism** of a group G is an isomorphism of G onto itself.

Notice that an automorphism is *onto*. The mapping $n \longrightarrow 2n$ is an isomorphism of $\langle J,+\rangle$ to itself which is not onto and thus does not qualify as an automorphism. A simple example of an automorphism is the mapping $a \longrightarrow a^3$ of a cyclic group of order 4. If the group is $\langle J_4,+_4\rangle$, then $0 \longrightarrow 0$, $1 \longrightarrow 3$, $2 \longrightarrow 2$, and $3 \longrightarrow 1$. That this is an automorphism is the result of direct calculation.

Theorem 3-6

Let $\langle G, \ \rangle$ be a group and $a \in G$. Then the mapping $I_a : x \longrightarrow a^{-1}xa$, for all $x \in G$, is an automorphism of G (called an *inner automorphism*).

The proof of this theorem is easy and is deferred to the exercises. Since the condition $a^{-1}xa = x$ is equivalent to $xa = ax$, an abelian group has only one inner automorphism—every I_a being the identity mapping. However, every group of order greater than 2 has at least one automorphism different from the identity mapping. For the non-abelian groups it is I_a for some a which fails to commute with every element of the group; for abelian groups in which $x \neq x^{-1}$ for some element it is the mapping $x \longrightarrow x^{-1}$; for groups such that $x = x^{-1}$ for every x, it is an appropriate interchange of two or more elements.

The fact that every subgroup of an abelian group is normal is closely related to the fact that the only inner automorphism of an abelian group is the identity mapping since normal subgroups can be characterized by the way inner automorphisms act on their elements.

Definition 3-7

A subgroup H of a group G is **invariant** in G iff $a^{-1}Ha \subset H$ for all $a \in G$.

Theorem 3-7

A subgroup H of a group G is invariant in G if, and only if, it is normal in G.

PROOF. Let H be an invariant subgroup of G and let $a \in G$. If $x \in aH$, then $x = ah = (aha^{-1})a$ for some $h \in H$. Since H is invariant, $aha^{-1} = (a^{-1})^{-1}ha^{-1} = h' \in H$. Thus $x = h'a \in Ha$, so $aH \subset Ha$. If $y \in Ha$, then $y = ka$ for some $k \in H$. So $y = a(a^{-1}ka) = ak^*$ for some $k^* \in H$. Thus $Ha \subset aH$, and we have proved that $aH = Ha$ whenever H is invariant. So H is normal if H is invariant.

Conversely, assume H is normal and let $a \in G$. Let $x \in a^{-1}Ha$. Thus $x = a^{-1}ha$ for some $h \in H$. Since H is normal $(aH = Ha)$, $ha = ah'$ for some $h' \in H$. Then $ax = ha = ah'$, so $x = h' \in H$. Thus $a^{-1}Ha \subset H$, and H is invariant. ||

Corollary 3-7*a*

A subgroup H of a group G is normal in G iff $Ha \subset aH$ for every $a \in G$.

PROOF. In the first part of the proof above, invariance of H implied $Ha \subset aH$ for all a. In the second part of the proof, the condition $aH = Ha$ could have been weakened to $Ha \subset aH$ to get the key statement $ha = ah'$. ||

These alternatives to the definition of normal subgroup are quite convenient since they are stated in terms of a subset relation rather than equality. In effect, what we have proved is that $aH = Ha$ will hold iff any one of the relations $a^{-1}Ha \subset H$, $aH \subset Ha$, or $Ha \subset aH$ holds. This cuts by half the work of proving normality in any situation.

The next theorem verifies that the process of making a group out of the left cosets of a subgroup (as was suggested in Exercises 3-24 and 1-84) can be accomplished if the subgroup is normal. This process will lead soon to the *fundamental theorem of group homomorphisms.*

Theorem 3-8

If H is a normal subgroup of G, if $x \in aH$, and if $y \in bH$, then (1) $xy \in (ab)H$, and (2) the collection of all left cosets of H in G is a group with respect to the operation $\square$ defined by

$$aH \square bH = (ab)H$$

PROOF. If $x \in aH$ and $y \in bH$, then there are elements $h,k \in H$

such that $x = ah$ and $y = bk$. Since H is normal, there is an element $g \in H$ such that $hb = bg$. Thus

$$xy = (ah)(bk) = a(hb)k = a(bg)k = (ab)(gk)$$

Since $gk \in H$, xy is an element of $(ab)H$.

The second part of the proof is left as an exercise. ||

If $x \in (ab)H$, then it can be represented as a product of elements from aH and bH, for if $h \in H$ is such that $x = (ab)h$, then

$$x = (ab)h = a(bh) = (ae)(bh)$$

Because of this, the calculation in the proof above can legitimately be rewritten as

$$(aH)(bH) = a(Hb)H = a(bH)H = (ab)(HH) = (ab)H$$

This sort of notational shortcut when computing with normal subgroups can be of some convenience.

Definition 3-8

If H is a normal subgroup of a group G and $G/H = \{aH \mid a \in G\}$ is the set of all left cosets of H in G, then $\langle G/H, \square\rangle$, with $\square$ as defined in Theorem 3-8, is called the **quotient group** (or **factor group**) of G modulo H.

Exercises 3-32 to 3-42

3-32. Show that the mapping $\beta : a \to a^{-1}$ is a group automorphism iff the group is abelian.

3-33. Show that the mapping $\alpha : n \to 2n$ is an isomorphism of $\langle J, +\rangle$ to itself.

3-34. The set $C = \{x \mid xa = ax \text{ for all } a \in G\}$ was shown in Exercise 2-25 to be a subgroup (called the center) of a group G. Show that C is normal in G.

3-35. Show that normality is not transitive by exhibiting a group G of order 8 and subgroups $H \subset K \subset G$ such that H is normal in K and K is normal in G but H is not normal in G. (See Exercise 3-31.)

3-36. Prove Theorem 3-6. (Prove I_a onto, one-to-one, and operation-preserving.)

☆***3-37***. Prove part 2 of Theorem 3-8.

3-38. Prove that the hypothesis that H be normal in Theorem 3-8 is essential to the conclusion.

3-39. Define the product AB of two subsets of a group G by

$$AB = \{ab \mid a \in A \text{ and } b \in B\}$$

a. Prove that $HH = H$ iff H is a subgroup of G.

☆b. Show that AB may not be a subgroup if A and B are subgroups.

3-40. Compute the inner automorphism I_a of the group G in each of the following cases:

a. $a = P$ in Table 3 of Appendix I

b. $a = R$ in the same group

c. $a = 2$ in Table 2 of Appendix I

3-41. Prove that the set $\mathfrak{A}(G)$ of all automorphisms of a group G is a group with composition of mappings as the operation.

3-42. Prove that the set of all inner automorphisms of G is a normal subgroup of $\mathfrak{A}(G)$.

3-D THE FUNDAMENTAL THEOREM

In this section the study of groups reaches the level of an important theorem relating normal subgroups to homomorphisms. This theorem is a basic piece of information for almost all further study of groups (as in Chapter 6). It also provides, in its proof, some valuable information for the study of two-operation systems—a study which will be undertaken in Chapter 4.

The fundamental theorem of group homomorphisms says essentially that there is a one-to-one correspondence between the normal subgroups of a group and the homomorphisms that can be defined on the group. In order to capture the important smaller pieces of information to be found in this theorem, we shall first consider the theorem in three parts and later assemble the parts.

Definition 3-9

Let α be a homomorphism from a group $\langle G, \circ \rangle$ onto a group $\langle \bar{G}, * \rangle$, and let $\bar{e}$ be the identity of $\bar{G}$. Then

$$K = \{x \mid x \in G \text{ and } x\alpha = \bar{e}\}$$

is called the **kernel** of α.

If there is some possibility that the kernels of two different homomorphisms might be confused, the notation K_α will be used for the kernel of α.

Observe that the kernel of α is a subset of G, not of $\bar{G}$. The kernel plays an important part in the relationship between homomorphisms and normal subgroups. It is, in fact, a normal subgroup.

Theorem 3-9*a*

The kernel K of a homomorphism α of group G onto $\bar{G}$ is a normal subgroup of G.

PROOF. To show that K is a subgroup of G, using Theorem 2-6, we observe that $e \in K$ since $e\alpha = \bar{e}$ (by Exercise 1-69) so $K \neq \varnothing$. If $x,y \in K$ then $x\alpha = y\alpha = \bar{e}$ and

$$(x \circ y^{-1})\alpha = x\alpha * y^{-1}\alpha = x\alpha * (y\alpha)^{-1} = \bar{e} * \bar{e}^{-1} = \bar{e}$$

Thus $x \circ y^{-1} \in K$ if $x,y \in K$, which implies that K is a subgroup of G.

To show that K is normal, we first prove that $x \in a \circ K$ iff $x\alpha = a\alpha$. If $x \in a \circ K$, then $x = a \circ k$ for some $k \in K$ (that is, for some k such that $k\alpha = \bar{e}$). Then

$$x\alpha = (a \circ k)\alpha = a\alpha * k\alpha = a\alpha * \bar{e} = a\alpha$$

Conversely, if $x\alpha = a\alpha$, then

$$(x^{-1} \circ a)\alpha = x^{-1}\alpha * a\alpha = (x\alpha)^{-1} * x\alpha = \bar{e}$$

This shows that $x^{-1} \circ a \in K$; thus $x \in a \circ K$ by Theorem 3-3. A similar calculation would show that $x \in K \circ a$ iff $x\alpha = a\alpha$.

Combining these results, $x \in a \circ K$ iff $x\alpha = a\alpha$ iff $x \in K \circ a$; thus $a \circ K = K \circ a$ for all $a \in G$, and K is normal in G by definition. ||

In proving that $x \in a \circ K$ iff $x\alpha = a\alpha$, we have established a rather significant property of homomorphisms. It has already been established that $x \in a \circ K$ iff $x \circ K = a \circ K$ and that the left cosets $a \circ K$ of a subgroup K form a partition of the group G (Theorem 3-3). It follows that every member of a given left coset of the kernel K maps into the same element of $\bar{G}$ under the homomorphism α, and there must be, then, a one-to-one correspondence between elements of $\bar{G}$ and left cosets of K in G.

Theorem 3-9*b*

If α is a homomorphism of G onto $\bar{G}$ with kernel K, then the factor group $\langle G/K, \square\rangle$ (see Theorem 3-8 and Definition 3-8) is isomorphic to $\langle \bar{G}, *\rangle$ under the mapping $\beta : a \circ K \rightarrow a\alpha$ defined for all $a \in G$.

DISCUSSION. In this theorem we go beyond the assertion that the correspondence $a \circ K \leftrightarrow a\alpha$ is one-to-one and claim that, whenever a group $\bar{G}$ is a homomorphic image of G, it is possible to find a system of subsets of G which is algebraically the same as $\bar{G}$ and dispense with $\bar{G}$ altogether. In other words, for all "practical" purposes, the only homomorphic images of a group G are factor groups of G.

PROOF. The observation that the mapping $\beta : a \circ K \rightarrow a\alpha$ is one-to-one between left cosets of K in G and elements of $\bar{G}$ has already been made. The mapping β is clearly onto $\bar{G}$ because α is a mapping of G onto $\bar{G}$. The only thing left to show is that β is operation-preserving:

$$\begin{aligned}[(a \circ K) \square (b \circ K)]\beta &= [(a \circ b) \circ K]\beta = (a \circ b)\alpha \\ &= a\alpha * b\alpha = (a \circ K)\beta * (b \circ K)\beta \quad \|\end{aligned}$$

We have now established that every homomorphism of a group determines a normal subgroup, which in turn determines a factor group which is a homomorphic image of the group (because it is isomorphic to a homomorphic image). The only open question is: Is it possible to begin with any normal subgroup and construct a factor group which is a homomorphic image of the group, or must one begin with the kernel of a homomorphism? This question can be answered by either of two assertions: (1) every factor group is a homomorphic image of the group, or (2) every normal subgroup is the kernel of some homomorphism.

Theorem 3-9c

Let H be any normal subgroup of a group $\langle G,\circ\rangle$. Then the mapping $\gamma : G \to G/H$ defined by

$$a\gamma = a \circ H \text{ for all } a \in G$$

is a homomorphism of G onto G/H with kernel H.

PROOF. The mapping γ makes each element of G correspond to the left coset determined by that element, so γ will be one-to-one iff $H = \{e\}$. But γ is definitely a mapping *onto* G/H since every left coset is determined by some element of G. Also γ is a homomorphism since

$$(a \circ b)\gamma = (a \circ b) \circ H = (a \circ H) \square (b \circ H) = a\gamma \square b\gamma$$

Since H is the identity of G/H and the elements a such that

$$a\gamma = a \circ H = H$$

are exactly those elements in H, the kernel of γ is H. ||

Finally, since we have now established that every homomorphism determines a normal subgroup (the kernel) and every normal subgroup determines a homomorphism which, in turn, has that normal subgroup as its kernel, we summarize these results in a compact form.

Theorem 3-9 (Fundamental theorem of group homomorphisms)

Every homomorphism of a group G determines a normal subgroup of G. Conversely, every normal subgroup of G determines a homomorphic image of G, and the factor groups of G modulo the normal subgroups are the only homomorphic images of G (up to isomorphism).

Figure 3-2 illustrates the relationship between the various groups, subgroups, and elements discussed in the three parts of Theorem 3-9. Note that $\alpha = \gamma\beta$ and $\gamma = \alpha\beta^{-1}$.

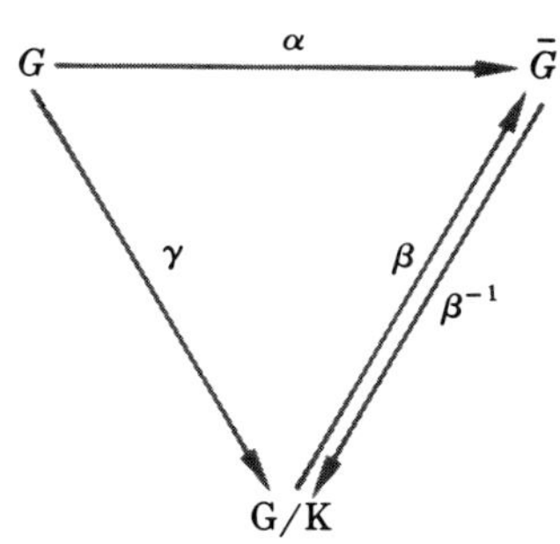

Figure 3-2 Diagram of the fundamental theorem.

Exercises 3-43 to 3-49

3-43. Let K be the kernel of homomorphism $\alpha : G \to \bar{G}$. Prove that $x \in K \circ a$ iff $x\alpha = a\alpha$.

3-44. Every factor group of a cyclic group is cyclic.

3-45. Discuss Theorem 3-9 in each of the following cases. (Identify the elements of Figure 3-2 in each case.)

a. G and $\bar{G}$ are isomorphic

b. $\bar{G} = \{\bar{e}\}$, a one-element group

☆c. K has index 2 in G

3-46. Discuss Theorem 3-9 if the group G is $\langle J, + \rangle$ and the subgroup is $\boldsymbol{H} = \{0, \pm 4, \pm 8, \pm 12, \ldots\}$. Make a table for $J/\boldsymbol{H}$.

3-47. Let α be a homomorphism of a group G onto $\bar{G}$ with kernel K. Prove that α is an isomorphism iff $K_\alpha = \{e\}$, where e is the identity of G.

3-48. Let α be a homomorphism of G onto $\bar{G}$. Let H be a normal subgroup of G. Prove that $\bar{H} = \{h\alpha \mid h \in H\}$ is a normal subgroup of $\bar{G}$.

3-49. Let α be a homomorphism of G onto $\bar{G}$. Let $\bar{H}$ be a subgroup of $\bar{G}$. Prove that $H = \{h \mid h\alpha \in \bar{H}\}$ is a subgroup of G which contains the kernel of α. Show also that if $\bar{H}$ is normal in $\bar{G}$, then H is normal in G.

▲

The subject of group theory is continued in Chapter 6, where we consider three topics which are direct developments from the theory of homomorphisms and normal subgroups discussed here. The fundamental theorem (3-9) is used there as an extremely valuable tool for obtaining several of the principal results.

The interested reader may proceed directly to Chapter 6 and defer Chapters 4 and 5 until later.

4 TWO-OPERATION SYSTEMS

The systems called groups include each of the usual number systems with the addition operation. This chapter considers these systems with both the addition and multiplication operations and considers also other systems which have an addition operation obeying the axioms of an abelian group and a multiplication operation with at least some of those properties. The two operations will be related by a distributive property. The particular kinds of systems to be considered are called rings, integral domains, and fields, and the progress from rings to fields involves annexing additional axioms about the multiplication operation.

4-A RINGS

A ring is a relatively weak two-operation system satisfying some, but not all, of the properties of the number systems.

Definition 4-1

A **ring** is an algebraic system $\langle A,+,\cdot\rangle$ with operations addition $(+)$ and multiplication $(\cdot)$ such that

R1: $\langle A,+\rangle$ is an abelian group
R2: Multiplication is associative in A
R3: Multiplication is distributive (left and right) over addition

It is convenient to adopt for the discussion of rings a number of notational devices used in the number systems. One should always interpret $a + b \cdot c$ (without parentheses) as $a + (b \cdot c)$. The symbol "0" (called *zero*) will be used for the identity for addition. The additive inverse of b will be denoted by ^-b or $-b$ and called the *negative* of b, and $a + (-b)$ will be written $a - b$.

It will also be convenient to use

$$na = a + a + \cdots + a \qquad n \text{ terms}$$

$$a^n = a \cdot a \cdot \cdots \cdot a \qquad n \text{ factors}$$

The expression na obeys the additive form of the rules of exponents proved in Theorem 2-10, and the expression a^n obeys the same rules for $n > 0$. The expression a^n may be meaningless in a ring if $n \leq 0$ since there is nothing in the definition of a ring to indicate that a multiplicative identity or multiplicative inverses exist. In the additive notation the group properties of the inverse $-a$ are (by Theorem 2-3):

$$-(-a) = a$$

$$-(a + b) = (-b) + (-a) = -b - a = -a - b$$

$$-(a - b) = b - a$$

The multiplicative properties of negatives (additive inverses) are exactly those of the number systems and can be proved for all rings. Of course, each number system R, Q, and J with the usual addition and multiplication is a ring.

Theorem 4-1

Let $\langle A,+,\ \rangle$ be a ring and $a,b \in A$. Then (1) $a0 = 0a = 0$, (2) $a(-b) = (-a)b = -(ab)$, and (3) $(-a)(-b) = ab$.

PROOF. *1.* To show $a0 = 0$, we calculate as follows:

$$a0 = a(0 + 0) = a0 + a0$$

$$a0 - a0 = (a0 + a0) - a0 = a0 + (a0 - a0)$$

$$0 = a0 + 0 = a0$$

2. Now $0 = a0 = a[b + (-b)] = ab + a(-b)$, and since inverses are unique in the group $\langle A,+\rangle$,

$$a(-b) = -(ab)$$

The remainder of the proof is deferred to the exercises. ||

Thus the properties of a ring are sufficient to make the zero element behave in multiplication just as the zero of the familiar number systems. This result is, of course, one of the advantages of the abstract approach to studying algebraic systems—it is possible to determine just what properties of a familiar system cause it to have certain other properties which are too often taken for granted.

Exercises 4-1 to 4-7

4-1. Write out symbolically in detail the axioms for a ring. (For example, property R1 involves five axioms.) Use the notational conventions introduced for identity and inverses.

4-2. Complete the proof of Theorem 4-1. [Show $0a = 0$, $(-a)b = -(ab)$, and $(-a)(-b) = ab$.]

☆*4-3.* In a ring $\langle A,+,\ \rangle$, $a(b - c) = ab - ac$ and $(b - c)a = ba - ca$.

4-4. In a ring $-(a - b) = b - a$.

4-5. Let $\langle A,+\rangle$ be a commutative group with identity 0. Let $a \cdot b = 0$ for all $a,b \in A$. Is $\langle A,+,\cdot\rangle$ a ring?

4-6. Let $\langle A,+\rangle$ be a commutative group. Let $a * b = a$ for all $a,b \in A$. Is $\langle A,+,*\rangle$ a ring?

4-7. a. Show that addition is not left-distributive over multiplication in $\langle J,+,\cdot\rangle$.

b. Assume that all the ring properties hold in J (which they do) and assume also that addition is distributive over multiplication. Show that $1 = 2$.

▲

The number system examples of rings are all too strong in that multiplication in each case is commutative and has an identity 1. Also, if m and n are nonzero numbers, then $m \cdot n \neq 0$. None of these may hold in general in a ring. A good example of a ring which satisfies almost no properties in addition to those given in Definition 4-1 is the ring $\langle T_2(E),+,\ \rangle$ of all 2×2 matrices of even integers

$$(E = \{0, \pm 2, \pm 4, \pm 6, \ldots\})$$

Addition and multiplication for matrices were defined in Definition 2-8. It is easy to find examples of matrices $M,N \in T_2(E)$ such that $MN \neq NM$ and to show that

$$I = \begin{bmatrix} 1 & 0 \\ 0 & 1 \end{bmatrix} \notin T_2(E)$$

is the only candidate for a multiplicative identity. Also the product of nonzero matrices can be the zero matrix (additive identity) since

$$\begin{bmatrix} 2 & 2 \\ 4 & 4 \end{bmatrix} \begin{bmatrix} 2 & -6 \\ -2 & 6 \end{bmatrix} = \begin{bmatrix} 0 & 0 \\ 0 & 0 \end{bmatrix}$$

Definition 4-2

A ring $\langle A,+,\cdot\rangle$ is called a **commutative ring** iff $\cdot$ is commutative in A. It is a **ring with unity** iff A contains an identity for multiplication.

The phrases "commutative" and "with unity" both refer to the multiplication operation in rings. (The addition operation has these properties in every ring.) It is convenient to use the combined phrase "commutative ring with unity" to indicate a ring with both of these additional properties. One should note that even a commutative ring with unity may still lack multiplicative inverses. The system of integers $\langle J, +, \cdot \rangle$ is a good example of this situation in which multiplicative inverses do not exist (except for 1 and -1), or, equivalently, equations $mx = n$ are not generally solvable. Other examples of commutative rings with unity are the "clock arithmetic" systems $\langle J_n, +_n, \cdot_n \rangle$. Consider the example of J_6 tabulated below.

$+_6$	0	1	2	3	4	5
0	0	1	2	3	4	5
1	1	2	3	4	5	0
2	2	3	4	5	0	1
3	3	4	5	0	1	2
4	4	5	0	1	2	3
5	5	0	1	2	3	4

$\cdot_6$	0	1	2	3	4	5
0	0	0	0	0	0	0
1	0	1	2	3	4	5
2	0	2	4	0	2	4
3	0	3	0	3	0	3
4	0	4	2	0	4	2
5	0	5	4	3	2	1

As in the example of $T_2(E)$ above, we again encounter the phenomenon $a \cdot b = 0$ when neither $a = 0$ nor $b = 0$. This does not happen in the usual number systems, including $\langle J, +, \cdot \rangle$, so there is still another property at work—one which does not necessarily depend on the existence of inverses.

Definition 4-3

An element $a \neq 0$ in a ring $\langle A, +, \quad \rangle$ is called a **zero-divisor** (or *divisor of zero*) iff there exists an element $b \neq 0$ in A such that $ab = 0$ or $ba = 0$. (Of course, b is a zero-divisor, too.) A commutative ring with unity and without zero-divisors is called an **integral domain.**

Thus in an integral domain, by definition, the equation $ab = 0$ implies that $a = 0$ or $b = 0$. The properties of integral domains will be considered in detail in Section 4-C.

It is characteristic of more advanced works on algebra to consider left zero-divisors and right zero-divisors. A left zero-divisor would be an $a \neq 0$ such that, for some $b \neq 0$, $ab = 0$. The element b would be a right zero-divisor. The ring $\langle J, +, \cdot \rangle$ is an integral domain. If E is the set of even integers, then $\langle E, +, \cdot \rangle$ (no unity), $\langle T_2(E), +, \quad \rangle$ (not commutative), and $\langle J_6, +_6, \cdot_6 \rangle$ (zero-divisors) are not integral domains.

Exercises 4-8 to 4-15

4-8. Exhibit $M,N \in T_2(E)$ such that $MN \neq NM$.

4-9. Let $\langle A,+,\ \rangle$ be a ring such that the additive group $\langle A,+\rangle$ is cyclic. Show that A is a commutative ring.

4-10. Show that $\langle P(X),\Delta,\cap\rangle$ is a commutative ring with unity. (See Theorem 2-8.)

4-11. Show that $\langle\{0\},+,\ \rangle$ is a commutative ring with unity.

☆**4-12.** Let $\langle A,+,\ \rangle$ be a ring with unity element e, where A contains at least two elements. Prove that $0 \neq e$.

4-13. Find all zero-divisors in J_8. Find all elements of J_8 which have multiplicative inverses. Do the same for J_{10}. Generalize, if possible, your conclusions.

4-14. For what positive integers n is J_n an integral domain?

4-15. Consider the following tables for $N = \{0,1,2,3\}$.

+	0	1	2	3
0	0	1	2	3
1	1	0	3	2
2	2	3	0	1
3	3	2	1	0

·	0	1	2	3
0	0	0	0	0
1	0	1	2	3
2	0	1	2	3
3	0	0	0	0

Show that N is the smallest noncommutative ring. (See Exercise 4-9.) Show that N has zero-divisors and does not have a unity. However, find a subring of N which is an integral domain. (See Definition 4-4 below.)

4-B SUBRINGS AND IDEALS

The theory of homomorphisms of rings (mappings which preserve both ring operations) is quite similar to the theory of homomorphisms of groups. In fact, it is possible to identify each ring homomorphism with a particular kind of subring in the same way that group homomorphisms are identified with normal subgroups. The subgroup and normal subgroup have their analogs for homomorphism theory in the concepts of the following definition.

Definition 4-4

Let $\langle A,+,\ \rangle$ be a ring. A subset B of A is a **subring** of A iff $\langle B,+,\ \rangle$ is a ring. A subring I of A is an **ideal** in A iff the product of an element of A and an element of I is always an element of I (that is, iff $x \in I$ and $a \in A$ imply $xa \in I$ and $ax \in I$).

Note that $B \subset A$ is a subring iff it is a ring with respect to the operations previously defined in A. In particular, the sum and product of two elements of B must be in B. The condition to be satisfied if B is an ideal is still stronger—B must be closed under multiplication if one of the elements is outside B. One might also consider left-ideals ($ax \in I$) and right-ideals ($xa \in I$) if he is interested in a more elaborate theory than that developed below. The set I of all multiples of 3 (or of any fixed integer) is a good example of an ideal in the ring of integers since I is a subgroup and, if $n \in J$ and $3k \in I$, then $n \cdot 3k = 3(nk) \in I$. Another similar example is the set

$$B = \{0,2,4\}$$

in J_6; this is also an ideal. On the other hand, the integers J, as a subset of the rational numbers Q, is a subring of Q which is not an ideal. Also the set of matrices of the form

$$\begin{bmatrix} m & 0 \\ 0 & 0 \end{bmatrix}$$

in $T_2(J)$ is a subring which is not an ideal, but $T_2(E)$ is an ideal in $T_2(J)$—where E is the set of all even integers.

Theorem 4-2

The nonempty subset B of a ring A is a subring of A if, and only if, $a - b$ and ab are in B for all $a,b \in B$.

PROOF. This theorem is the analog of Theorem 2-6, the basic subgroup theorem, which is used to prove this result.

If B is a subring of A, then B is a ring and must contain additive inverses, sums, and products of its elements.

Conversely, if $B \neq \varnothing$ and contains $a - b = a + (-b)$ for all $a,b \in B$, then $\langle B,+\rangle$ is a subgroup of $\langle A,+\rangle$ by Theorem 2-6. Since associativity of multiplication and distributivity are hereditary properties, the fact that the product of every two elements of B is in B is all that is needed to make B a ring. ||

Actually it is not necessary to show that a subset I of a ring A is a subring of A before showing that I is an ideal in A since the condition that $xa \in I$ for all $x \in I$ and $a \in A$ implies that $xa \in I$ if both x and a are in I. This comment makes the next theorem possible.

Theorem 4-3

A nonempty subset I of a ring A is an ideal in A if, and only if, $x - y \in I$ for all $x,y \in I$ and xa and ax are in I for every $x \in I$ and every $a \in A$.

PROOF. This is an almost trivial simplication of the previous theorem and Definition 4-4. If I is an ideal, the two conditions must hold. Conversely, if the first condition holds, I is a subgroup of the additive group of A; the second condition is the condition that a subring be an ideal, and that condition implies that $xy \in I$ if x and y are both in I. ||

Theorems 4-2 and 4-3 are useful in checking to see if subsets of a ring are subrings and ideals of the ring. Part of the check, according to the theorems, is to see that the sets are nonempty.

It was mentioned earlier that the ideals are to ring theory what the normal subgroups are to group theory. This assertion is amplified below as we make a ring out of the left cosets of the additive group of an ideal. Since an ideal I in a ring $\langle A,+,\ \rangle$ is a subring and therefore a subgroup of the additive group, which is abelian, I is a normal subgroup of the additive group $\langle A,+\rangle$. Thus it is possible to talk about left cosets $a + I$ and assert that left and right cosets of I determined by $a \in A$ are equal. Also the addition of cosets by the rule

$$(a + I) + (b + I) = (a + b) + I$$

is well-defined as was proved in Theorem 3-8. There is a strong temptation to define a multiplication of these sets by

$$(a + I)(b + I) = ab + I$$

The condition that $xa, xb, bx, ax \in I$ if $x \in I$ and the calculation

$$(a + I)(b + I) = ab + Ib + aI + II \subset ab + I + I + I = ab + I$$

makes this multiplication reasonable. (The subset relation in the last statement is necessitated by the fact that all that was assumed about xb for $x \in I$ was that $xb \in I$. It cannot be known if $\{xb \mid x \in I\}$ is equal to I without specifying a particular I, and this may not be true for certain I.)

Definition 4-5

Let I be an ideal in a ring A and let $a \in A$. Then the set

$$a + I = \{a + k \mid k \in I\}$$

is called the **coset** (or *residue class*) determined by a and I. The set A/I of all cosets of I in A with operations $+$ and $\cdot$ defined by

$$(a + I) + (b + I) = (a + b) + I$$

and

$$(a + I) \cdot (b + I) = ab + I$$

is called the **quotient ring** (or *residue class ring*) of A modulo I.

To justify the use of the word "ring" in regard to A/I, it is necessary to prove the following theorem.

Theorem 4-4

Let I be an ideal in a ring A. Then $\langle A/I, +, \cdot \rangle$ is a ring.

Exercises 4-16 to 4-26

☆4-16. Show that multiplication is well-defined in A/I by showing that $x \in a + I$ and $y \in b + I$ imply $xy \in ab + I$.

4-17. Prove Theorem 4-4.

4-18. Show that

$$\left\{\begin{bmatrix} m & 0 \\ 0 & 0 \end{bmatrix} \mid m \in J\right\}$$

is a subring of $T_2(J)$ which is not an ideal.

4-19. Show that $T_2(E)$ is an ideal in $T_2(J)$.

4-20. Show that Q is a subring of R which is not an ideal.

4-21. Find all ideals in J_8.

4-22. Show that A and $\{0\}$ are always ideals in a ring A. (Ideals other than these are called *proper ideals.*)

4-23. (Principal ideals) Let A be a commutative ring and let $a \in A$. Then the set $P_a = \{xa \mid x \in A\}$ is an ideal in A, called the *principal ideal* generated by a.

4-24. The system $\langle P(X), \Delta, \cap \rangle$ is a commutative ring with unity (Exercise 4-10). Let $Y \subset X$. Show that $P(Y)$ is an ideal in $P(X)$.

4-25. If I and K are ideals in a ring A, then $I \cap K$ is an ideal in A.

4-26. If I and K are ideals in a ring A, then

$$I + K = \{j + k \mid j \in I \text{ and } k \in K\}$$

is an ideal in A.

Definition 4-6

Let $\langle A, +, \ \rangle$ and $\langle B, \oplus, \cdot \rangle$ be rings. A **homomorphism** (or *ring homomorphism*) of A to B is a mapping $\gamma : A \to B$ such that

$$(a + b)\gamma = a\gamma \oplus b\gamma \qquad \text{and} \qquad (ab)\gamma = a\gamma \cdot b\gamma$$

for all $a,b \in A$. If γ is a one-to-one mapping, it is called an **isomorphism.**

Thus a mapping between rings is a homomorphism iff it preserves both operations. The theory of ring homomorphisms is entirely analogous to that of group homomorphisms and centers around two results.

Theorem 4-5

If $\gamma : A \to B$ is a ring homomorphism, then rn $(\gamma) \subset B$ is a subring of B.

DISCUSSION. This theorem enables us to speak of a homomorphism from A *onto* B without loss of generality. Actually it is true that if $\langle B, \oplus, \cdot \rangle$ is a two-operation system which is not a ring and $\gamma : A \to B$ where A is a ring and γ preserves operations, then rn (γ) is a ring with respect to the operations of B. All of this follows from Theorems 1-3 to 1-5—except for the distributive laws in rn (γ). These follow from Exercise 1-71.

The relationship between homomorphisms and ideals in ring theory is the same as that between homomorphisms and normal subgroups in group theory. In fact, since an ideal is a normal subgroup of the additive group, much of the group homomorphism theory can be used to simplify the proof.

Theorem 4-6 (Fundamental theorem of ring homomorphisms)

Let γ be a homomorphism of ring A onto ring B and let z be the zero of B. Then the *kernel* of γ, $I = \{a \in A \mid a\gamma = z\}$, is an ideal in A, and B is isomorphic to the quotient ring A/I. Conversely, if K is any ideal in A, then A/K is a homomorphic image of A with kernel K.

Exercises 4-27 to 4-31

4-27. Prove Theorem 4-6. (Follow an outline similar to that of Theorem 3-9, and use the results of Theorems 3-9*a*, *b*, and *c* where possible.) Does Figure 3-2 apply to this situation?

4-28. Define a homomorphism of $\langle J, +, \cdot \rangle$ onto $\langle J_4, +_4, \cdot_4 \rangle$. What is the kernel of the homomorphism?

4-29. Define the product MN of subsets M and N of a ring A by $MN = \{ab \mid a \in M \text{ and } b \in N\}$. Let I be an ideal in A. Show that $(a + I)(b + I) \subset ab + I$ but that equality may not hold. (There is an example using $A = J$ and $I =$ the multiples of 5.)

4-30. Let γ be a homomorphism from the ring A onto the ring B and let B_1 be a subring (ideal) of B. Show that

$$B_1\gamma^{-1} = \{x \in A \mid x\gamma \in B_1\}$$

is a subring (ideal) of A.

4-31. Let A be a ring with unity element e. Let I be an ideal of A such that $e \in I$. Show that $I = A$.

4-C INTEGRAL DOMAINS

As stated in Definition 4-3, an integral domain is a commutative ring with a unity element and with no zero-divisors. Other authors may give a slightly different definition of integral domain; in particular, they may omit commutativity, unity, or both. Our definition, however, appears to be the most usual one. The properties of divisibility and order are most conveniently developed within the context of an integral domain, and it is these two topics which will be considered next. The system of integers $\langle J,+,\cdot\rangle$ is the easiest and most important example of the theory at this point.

The integers are an example of a system in which cancellation laws hold for multiplication although multiplicative inverses do not exist (except for 1 and -1). Although a proof of cancellation laws for multiplication involving multiplying both sides of $ab = ac$ by a^{-1} $(a \neq 0)$ would assume more than is given in an integral domain, it is still true—because of the lack of zero-divisors—that $b = c$.

Theorem 4-7

If $\langle D,+,\ \rangle$ is an integral domain and $a,b,c \in D$ such that $a \neq 0$, and if $ab = ac$, then $b = c$.

PROOF. If $ab = ac$, then $ab - ac = a(b - c) = 0$. But D is an integral domain, so the assertions $a(b - c) = 0$ and $a \neq 0$ combine to imply that $b - c = 0$. So $b = c$. ||

A converse of this theorem, to the effect that a commutative ring with unity in which the cancellation law holds must be an integral domain, is also true. (See Exercise 4-32.)

Having established the cancellation law for nonzero elements, the attention shifts to a study of the important topic of divisibility in integral domains.

Definition 4-7

Let D be an integral domain and $a,b \in D$. Then a **divides** b (a is a *factor* of b; b is a *multiple* of a) iff there exists $c \in D$ such that $ac = b$. If a divides b, write "$a \mid b$," and if a does not divide b, write "$a \nmid b$."

One should note that no division operation is actually involved in the definition above. The word "divides" is defined strictly in terms of the multiplication operation in D, and no mention of a division operation will be made in any proofs involving this concept. This form of the definition permits the use of the zero element of D in places which would be impossible in a division operation. Thus $0 \mid 0$ is a meaningful (and true) statement about D, but $0/0$ has not been (and will not be) given meaning.

Definition 4-8

If $x \mid y$ and $y \mid x$, then x and y are called **associates.** An associate of the unity element e is called a **unit.**

In the integers, for example, the following statements are true: $2 \mid 14$, $3 \nmid 14$, $15 \mid (-75)$, $15 \nmid 85$, -1 is a unit, 5 and -5 are associates. In fact, the only associates of $n \in J$ are n and $-n$.

A couple of simple results have been overly dignified by being listed in the following theorem.

Theorem 4-8

Let D be an integral domain with unity element e. Let $a,b \in D$. Then (1) $a \mid a$, (2) $a \mid 0$, (3) $e \mid a$, and (4) $0 \mid b$ iff $b = 0$.

PROOF. (1) and (3) $ae = a$; (2) $a0 = 0$; (4) $0 \mid b$ iff there exists c such that $0c = b$, but then $b = 0$. ||

The concept of divisibility is of no consequence in the study of the rational and real number systems, where every nonzero element has an inverse. In those systems every nonzero number divides every number, and every nonzero number is an associate of every other nonzero number. The basic properties regarding the ideas of divisibility, associates, and units are contained in the following theorem, the proof of which is deferred to the exercises.

Theorem 4-9

Let D be an integral domain and $a,b,c,x \in D$. Then

1. If $a \mid b$ and $b \mid c$, then $a \mid c$

2. The element x is a unit of D iff x^{-1} exists in D

3. The element x is a unit of D iff x divides every element of D

4. The elements a and b are associates iff there exists a unit $x \in D$ such that $ax = b$.

Exercises 4-32 to 4-39

4-32. Let D be a commutative ring with unity such that the cancellation law ($ab = ac$ and $a \neq 0$ imply $b = c$) holds for all elements of D. Show that D is an integral domain.

4-33. (For calculus students) Show that the set $C[0,1]$ of all continuous functions on $[0,1]$ is a ring with respect to the usual addition and multiplication of functions. Show that $C[0,1]$ is not an integral domain.

4-34. If A is an integral domain, then any subring of A which contains the unity is an integral domain.

4-35. Prove Theorem 4-9.

4-36. The relation ρ defined by "$a \rho b$ iff a and b are associates" is an equivalence relation in an integral domain.

4-37. Show that Theorem 4-9 implies that if every nonzero element of D has an inverse, then $a \mid b$ for every $a,b \in D \sim \{0\}$.

4-38. Let $a,b,c \in D$, an integral domain. Consider the six propositions which can be formed from the three hypotheses

1. $a \mid b$ and $a \mid c$

2. $a \mid b$ and $a \nmid c$

3. $a \nmid b$ and $a \nmid c$

and the two conclusions

4. $a \mid (b + c)$

5. $a \mid bc$

Which of the six propositions are theorems about D? Which are sometimes true if $D = J$? Which are never true if $D = J$?

4-39. The product of two units in D is a unit in D.

Definition 4-9

An element p of an integral domain D is called **prime** iff p is not 0 or a unit and the only factors of p are the associates of p and the units. An element $c \in D$ is **composite** iff c is not 0, a unit, or a prime.

This definition partitions the integral domain into four classes (or fewer if one or more is empty): the first contains 0 alone, the second contains the units, the third the primes, and the fourth all remaining elements (the composites). Examples of primes in J, where the associates of an element x are x and $-x$, are 2, -17, 101, and -211. Some composites are $14 = 2 \cdot 7, 500 = 10 \cdot 50$, and $-1{,}001 = (-77) \cdot 13$. Observe that negative numbers in J can be primes in the sense of the definition above. This definition is thus more general than the one given earlier (Definition 2-7).

Exercises 4-40 to 4-48

4-40. List all the primes in J between 50 and 100.

4-41. What are the primes and composites in
a. $\langle Q,+,\cdot\rangle$ b. $\langle J_7,+_7,\cdot_7\rangle$

4-42. Find 10 consecutive integers between 100 and 150 all of which are composite.

4-43. Find a prime $p \in J$ such that $p + 10n$ is also prime for

$$n = 1,2,3,4$$

4-44 to 4-46. The *characteristic* of an integral domain D is the order of the subgroup $[e]$ of $\langle D,+\rangle$—that is, it is the smallest integer n such that $ne = 0$. (Let D have characteristic 0 if $[e]$ has infinite order.)

4-44. Every nonzero element of an integral domain has additive order equal to the characteristic.

4-45. The characteristic of an integral domain is not a composite integer.

☆*4-46.* In an integral domain of characteristic p, a prime,

$$(a + b)^p = a^p + b^p$$

for all a and b.

4-47. Prove that $\langle J_n, +_n, \cdot_n\rangle$ is an integral domain if, and only if, n is a prime.

4-48. Prove that Definitions 2-7 and 4-9 are equivalent for positive elements of J.

4-D ORDERED INTEGRAL DOMAINS

This section on order is devoted to the simple properties of the "less than" order relation in the slightly abstract setting of an integral domain. What is done here will be applicable to the integers, rational numbers, and real numbers. Very few of the theorems will be proved since the proofs are all reasonably easy.

Convention

It will be assumed that every system mentioned in the rest of this chapter contains at least two elements.

This convention makes it possible to avoid the complications of talking about $\{0\}$ as an integral domain, and Exercise 4-12 assures us that $0 \neq e$, the unity element, if the integral domain has two or more elements.

Definition 4-10

Let $\langle D,+,\ \rangle$ be an integral domain. A subset P of D is a **positive set** iff

P1: P is closed with respect to addition and multiplication.
P2: If $c \in D$, then exactly one of the following statements holds: $c \in P$, $-c \in P$, $c = 0$.

The members of P are called **positive elements.**

The positive set can be used to order the integral domain in the following way.

Definition 4-11

Let $a,b \in D$. We shall say that a is **less than** b (and write "$a < b$") iff $b - a \in P$. (We shall also write "$b > a$" to mean $a < b$.) The system $\langle D,+, \quad ;<\rangle$ is called an **ordered integral domain.**

The condition P2 of Definition 4-10, which states that $b - a \in P$, $a - b \in P$, or $a - b = 0$ for any $a,b \in D$, now gives us the *trichotomy* property of the relation $<$:

$$a < b \qquad b < a \qquad \text{or} \qquad a = b$$

A couple of elementary facts about the order relation are included in the following theorem.

Theorem 4-10

Let D be an ordered integral domain with positive set P. Then, for $a,b,c \in D$,

1. $a > 0$ iff $a \in P$.

2. If $a < b$ and $b < c$, then $a < c$.

PROOF. *1.* $a > 0$ iff $a - 0 = a \in P$.

2. If $a < b$ and $b < c$, then $b - a \in P$ and $c - b \in P$. But P is closed under addition (P1), so

$$(b - a) + (c - b) = c - a \in P$$

Thus $a < c$. ||

Theorem 4-11

If $a,b,c \in D$ and if $a < b$, then $a + c < b + c$.

Theorem 4-12

If $a,b,c \in D$, if $a < b$, and if $c > 0$, then $ac < bc$. If $c < 0$, then $ac > bc$.

PROOF. By hypothesis in the first case, $b - a \in P$ and $c \in P$. Thus $(b - a)c = bc - ac$ is an element of P. Thus $ac < bc$.

If $c < 0$, then $0 - c = -c \in P$. Thus

$$(b - a)(-c) = b(-c) - a(-c) = ac - bc$$

is in P; so $ac > bc$. ||

Theorem 4-13

If $a < b$ and $c < d$, then $a + c < b + d$.

The three theorems above are the principal results concerning the relationship of the "$<$" relation to the operations in D. The questions as yet unanswered are those concerning the properties of the unity element and inverses (if they exist) in regard to the "$<$" relation.

Theorem 4-14

If $a \neq 0$, then $a^2 > 0$.

Corollary 4-14*a*

The unity element e of an integral domain is positive.

The corollary above is truly a corollary since $e \neq 0$, so that $e^2 = e$ is positive. This result says something interesting about the integral domain J of integers. Since 1 is positive and every positive integer n can be written as $n1 = 1 + 1 + \cdots + 1$, the positive integers are the only subset of J which can be used as a positive set in the sense of Definition 4-10. A similar statement about the rational numbers—that Q_+ is the only possible positive set in Q—follows from the theorems about the relationships of inverses to the order relation.

Theorem 4-15

If $a > e$, then $a^2 > a$. If $0 < a < e$, then $a^2 < a$.

Theorem 4-16

1. If $a < b$, then $-a > -b$.

2. If a^{-1} exists in D, then $a^{-1} > 0$ iff $a > 0$.

3. If a^{-1} exists in D and $a > e$, then $a^{-1} < e$.

4. If a^{-1} and b^{-1} exist in D and if $0 < a < b$, then $b^{-1} < a^{-1}$.

Exercises 4-49 to 4-57

4-49. Show that the set $\{a + bi \mid a > 0\}$ is not a positive set in the set C of complex numbers. (Two properties fail.)

4-50. Prove Theorem 4-11.

4-51. Show that $a < b$ and $c < d$ does not imply $ac < bd$ in J. In what cases will $a < b$ and $c < d$ imply $ac < bd$ in an integral domain?

4-52. Prove Theorem 4-13 using previous theorems only.

4-53. Prove Theorem 4-14.

4-54. Prove Theorem 4-15.

4-55. Prove Theorem 4-16.

4-56. State and prove the theorems concerning the order relation between a^{-1} and b^{-1} if $a < 0 < b$ and if $a < b < 0$.

4-57. Let D be an ordered integral domain with unity e. Show that $[e] \subset D$ is ring isomorphic to $\langle J,+,\cdot\rangle$ under the correspondence $\alpha : ne \to n$. Show that this isomorphism preserves order (that is, if $a,b \in D$ and $a < b$ in D, then $a\alpha < b\alpha$ in J).

▲

Exercise 4-57 yields the characterization of the integers which was promised in Chapter 1. Since $\langle [e],+,\ \rangle$ in D is isomorphic to $\langle J,+,\cdot\rangle$ and order is preserved, the system of integers can be characterized as the *smallest ordered integral domain*—in the sense that every ordered integral domain contains a subdomain (namely $[e]$) isomorphic to J.

This idea can be made more precise by using the terminology of the following definition.

Definition 4-12

A system $\mathcal{B}$ is an **extension** of system $\mathcal{A}$ iff $\mathcal{B}$ contains a subsystem isomorphic to $\mathcal{A}$. System $\mathcal{A}$ is **embedded** in $\mathcal{B}$ iff $\mathcal{B}$ is an extension of $\mathcal{A}$.

Thus Exercise 4-57 says that J is embedded in every ordered integral domain. The principal result of the next section is another embedding theorem in which we shall attempt to embed every integral domain in a system in which every nonzero element has a multiplicative inverse.

Before passing from the integers to topics which pertain to the rational, real, and complex numbers, we shall pause to list a sequence of theorems about the integers J—a sequence which terminates in the unique factorization theorem. The proofs of these will not be given here. However, similar proofs will be given in the next chapter for the corresponding results on polynomials, and the reader will be asked then to prove these results.

This sequence of results actually begins with the axioms for the integers S, A, C, D, I, N, and V$^+$ of Section 1-F, with the equivalence of mathematical induction (axioms N and N′) and the well-ordering property (Exercise 1-75), and with the division algorithm (Theorem 1-6).

Definition 4-13

A **greatest common divisor** of two integers a and b is a positive number d ($d = \gcd\{a,b\}$) such that $d \mid a$ and $d \mid b$, and if c is an integer which divides a and b then c | d.

Theorem 4-17

If a and b are integers, at least one of which is not zero, then

$$d = \gcd\{a,b\}$$

exists, is unique, and can be expressed as

$$d = ax + by$$

where x and y are integers.

Definition 4-14

The integers a and b are **relatively prime** iff gcd $\{a,b\} = 1$.

Theorem 4-18

If a and b are relatively prime integers, if c is an integer, and if $a \mid bc$, then $a \mid c$.

Corollary 4-18a

If p is a prime in J and if $p \mid a_1a_2 \cdots a_n$, then $p \mid a_i$ for some i.

The next theorem is usually called the *unique factorization theorem* or the *fundamental theorem of arithmetic.*

Theorem 4-19

Every positive composite integer can be written as a product of positive primes—uniquely except for the order of the factors.

4-E FIELDS

Having considered those properties of two-operation systems common to all of the integers, rational numbers, real numbers, and (except for order) the complex numbers, we turn to the study of those systems which, unlike the integers, include a multiplicative inverse for every element except the zero. In this regard we consider only commutative systems and leave until the exercises an example of a noncommutative system which has inverses for every element except 0.

Definition 4-15

A **field** is a system $\langle F,+,\ \rangle$ (consisting of at least two elements) which is a commutative ring with unity in which every nonzero element has a multiplicative inverse.

In the sense of this definition the rational, real, and complex number systems are all fields. On a much smaller scale, the system $\langle J_2,+_2,\cdot_2\rangle$ is a field. The properties of integral domains are at our disposal for the study of fields since every field is an integral domain—which is the next theorem.

Theorem 4-20

Every field is an integral domain.

PROOF. Since a field is a commutative ring with unity, the only thing to be proved is that a field has no zero-divisors. This follows from the fact of inverses.

Let F be a field and $a,b \in F$ such that $a \neq 0$. Then if $ab = 0$, since a^{-1} exists in F,

$$a^{-1}(ab) = a^{-1}0$$

Thus $b = 0$. So if a product is zero, some factor must be zero. ||

Another equally uncomplicated proof of this theorem can be given using Exercise 4-32, which says that every commutative ring with unity

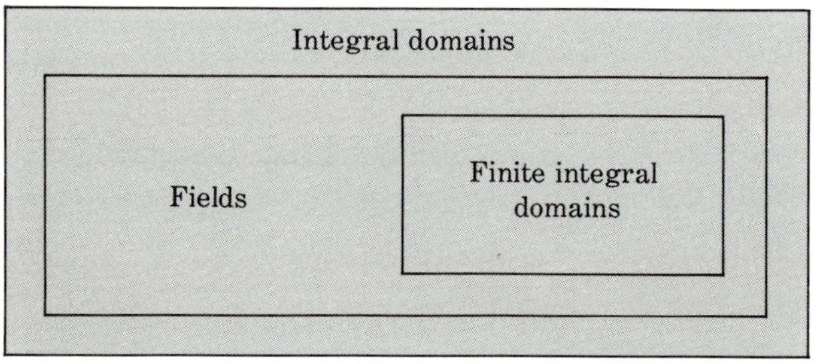

Figure 4-1 Relation of integral domains to fields.

in which the multiplicative cancellation law holds for nonzero elements is an integral domain. In a field F, if $a \neq 0$ and $ab = ac$, then a^{-1} exists and

$$a^{-1}(ab) = a^{-1}(ac)$$

so $b = c$, and F is an integral domain.

The converse of this statement is false. The system $\langle J,+,\cdot\rangle$ of integers is an integral domain which is not a field. However, fields and integral domains are equivalent concepts in the case of finite systems. (See Figure 4-1.)

Theorem 4-21

Every finite integral domain is a field.

The proof of this theorem follows the pattern, established in proving Theorem 2-7 and applicable to Exercise 2-15, of considering all the products of a fixed element of the system with the other elements. Therefore, the proof is deferred to the exercises.

Corollary 4-21*a*

The system $\langle J_n, +_n, \cdot_n \rangle$ is a field iff n is a prime.

PROOF. Exercise 4-47 and Theorem 4-21. ||

Exercises 4-58 to 4-63

4-58. Prove Theorem 4-21.

4-59 to 4-62. (***Quaternions***) Consider the set of 2×2 matrices with complex number entries of the form

$$\begin{bmatrix} a + bi & c + di \\ -c + di & a - bi \end{bmatrix} \qquad a,b,c,d \in R$$

with the usual addition and multiplication of matrices (see Definition 2-8). This system is called the system of *quaternions.*

4-59. The system of quaternions is a ring with unity.

4-60. Every quaternion has a multiplicative inverse in the system unless $a = b = c = d = 0$.

4-61. The ring of quaternions is not commutative. (A ring with unity and inverses for every nonzero element is called a *division ring.* Thus a field is a commutative division ring.)

4-62. Consider the set $K = \{a + bi + cj + dk \mid a,b,c,d \in R\}$ with addition and multiplication defined as for polynomials subject to the rules $i^2 = j^2 = k^2 = -1$, $ij = k$, $jk = i$, $ki = j$.

a. Show that $ji = -k$, $kj = -i$, and $ik = -j$.

☆b. Show that K is isomorphic to the quaternions.

c. Show that K (hence the quaternions) contains subsystems isomorphic to the real and complex numbers.

4-63. Let $A = \{0,1\}$. Consider the system $\langle A,+,\cdot\rangle$ with operations defined by the following tables.

$+$	0	1		$\cdot$	0	1
0	0	1		0	0	1
1	1	0		1	0	1

$\langle A,+\rangle$ is clearly an abelian group, and $A \sim \{0\} = \{1\}$ is just as clearly an abelian group with respect to multiplication. Show that multiplication is left-distributive but not right-distributive over addition. (See the articles, "A Remark Concerning the Definition of a Field," by A. H. Lightstone, January, 1964, and "An Additional Remark Concerning the Definition of a Field," by J. J. Malone, Jr., March, 1965, in *Mathematics Magazine* for the importance and uniqueness of this system.)

▲

We have shown that all fields are integral domains and that some integral domains are not fields. We are going to go a step farther and show that, although an integral domain may not be a field, it can always be embedded in one. Thus every integral domain is, in this sense, part of a field. The model for this construction is the relationship between the integers and the rational numbers.

The rational numbers are usually denoted as pairs of integers a/b in fraction form. Sometimes different pairs are intended to stand for the same rational number, as in $2/3 = 8/12$. In coming to grips with these ideas and in copying the form of addition and multiplication of fractions, we obtain a method for constructing a field from any integral domain.

Theorem 4-22

Every integral domain can be embedded in a field (called its *field of quotients*).

PROOF. Let $\langle D,+,\ \rangle$ be an integral domain with unity e and let

$$E = \{(a,b) \mid a,b \in D \text{ and } b \neq 0\}$$

Define a relation ρ in E by

$$(a,b)\ \rho\ (c,d) \text{ iff } ad = bc$$

The relation ρ is an equivalence relation in E because:

r. $(a,b)\ \rho\ (a,b)$ since $ab = ba$ in D.

s. If $(a,b)\ \rho\ (c,d)$, then $ad = bc$; so $cb = da$, which implies that $(c,d)\ \rho\ (a,b)$.

t. If $(a,b)\ \rho\ (c,d)$ and $(c,d)\ \rho\ (g,f)$, then $ad = bc$ and $cf = dg$. Since b, d, and f are not zero, if any one of a, c, and g is zero, then the others are, too, because D has no zero-divisors. But if $a = 0 = g$, then $(a,b)\ \rho\ (g,f)$. If none of these three is zero, then

$$(ad)(cf) = (bc)(dg)$$

and we may cancel c and d to get $af = bg$, which implies $(a,b)\ \rho\ (g,f)$.

We now know that ρ is an equivalence relation in E, and we have also observed that $(a,b)\ \rho\ (0,d)$ iff $a = 0$.

We shall denote the equivalence classes in E by $E(a,b)$—

$$E(a,b) = \{(x,y) \mid (x,y)\ \rho\ (a,b)\}$$

and the partition of E induced by ρ we shall denote by F. The set

$$F = \{E(a,b) \mid (a,b) \in E\}$$

is the set which we wish to make into a field by appropriate definitions of operations. (We have already an idea of "equality of fractions" from the relation ρ.) Define

$$E(a,b) + E(c,d) = E(ad + bc, bd)$$

and

$$E(a,b) \cdot E(c,d) = E(ac, bd)$$

We must first show that these operations are truly operations on the equivalence classes and do not depend on the member of the class used to denote that class. To do this for addition, let $(x,y) \in E(a,b)$. We wish to show that $(xd + yc, yd) \in E(ad + bc, bd)$. This will be true iff

$$(xd + yc)bd = yd(ad + bc)$$

iff

$$xdb + ycb = yad + ybc$$

But we are given $xb = ya$. Therefore the previous equality holds iff

$$(ya)d + cyb = yad + ycb$$

The proof for the multiplication definition is similar.

We now need to show that $\langle F,+,\cdot\rangle$ is a field. We omit the calculations for the associative, commutative, and distributive laws, which follow from the corresponding properties of D.

We have already noted that $E(a,b) = E(0,e)$ iff $a = 0$, and $E(0,e)$ is the zero class in F, for

$$E(x,y) + E(0,e) = E(xe + y0,ye) = E(x,y)$$

The negative of $E(a,b)$ is $E(-a,b)$, as can be verified by a quick calculation. Similarly $E(e,e)$ is the unity of F and $E(b,a)$ is the inverse of $E(a,b)$ $(a \neq 0)$. Thus F is a field.

To show that D is embedded in F, let

$$D' = \{E(a,e) \mid a \in D\}$$

Then the mapping $\alpha : a \rightarrow E(a,e)$ is a ring isomorphism of D onto D'. $\|$

If D is an ordered integral domain in Theorem 4-22 above, then the order relation can also be extended to the field F constructed from D.

Theorem 4-23

Let $\langle D,+,\cdot\,;<\rangle$ be an ordered integral domain. The set

$$P' = \{E(a,b) \mid ab > 0 \text{ in } D\}$$

is a positive set in the field F of quotients of D, and P' defines an order relation $\prec$ in F such that $E(a,e) \prec E(b,e)$ iff $a < b$ in D.

PROOF. The first thing to show is that if $(c,d) \in E(a,b)$, then $cd > 0$ iff $ab > 0$. But $(c,d) \in E(a,b)$ implies that $ad = bc$. If $ab > 0$, then $a > 0$ and $b > 0$, or $a < 0$ and $b < 0$. If $d > 0$ and $c < 0$ (or vice versa), then exactly one of ad and bc would be positive, and ad could not equal bc. Thus $d > 0$ and $c > 0$, or $d < 0$ and $c < 0$. In either case, $cd > 0$. Thus the definition of P' is independent of the element used to determine the equivalence class $E(a,b)$.

To show that P' is a positive set in F, let $E(a,b)$ and $E(c,d)$ be elements of P' such that $b > 0$ and $d > 0$ in D. Then $a > 0$ and $c > 0$ as well. Hence $ad + bc > 0$ and $bd > 0$. Thus

$$E(ad + bc,bd) = E(a,b) + E(c,d)$$

is in P', since $(ad + bc)bd$ is positive in D. In the same manner $E(a,b) \cdot E\,(c,d)$ can be shown to belong to P'. The trichotomy property, $E(a,b) \in P'$, $E(a,b) = E(0,b)$, or $-E(a,b) \in P'$, follows immediately since $ab > 0$, $ab = 0$, or $ab < 0$ in D.

Since $e > 0$ in D, $E(a,e) \prec E(b,e)$ iff

$$E(b,e) - E(a,e) = E(b - a,e)$$

is in P' iff $(b - a)e = b - a$ is positive in D iff $a < b$ in D. $\|$

Thus the order relation in the integers can be extended to an order relation in the rational numbers which preserves the order of the integers (thought of as a system embedded in the rationals).

We are now in a position to characterize the rational numbers in a manner similar to that characterization given to the integers in the last section. Let e be the unity element of an ordered field F. Then the mapping $\beta : Q \to F$ defined by $(m/n)\beta = (me)(ne)^{-1}$ with $n \neq 0$ is an isomorphism of Q onto the subfield $F' = \{(me)(ne)^{-1} \mid m,n \in J$ and $n \neq 0\}$ which preserves order. Thus the system Q of rational numbers is the smallest ordered field—in the sense that Q is embedded in every ordered field.

Exercises 4-64 to 4-67

4-64. Fill in the following gaps in the proof of Theorem 4-22.

a. Prove that the multiplication operation defined in F depends only on the classes and not on the ordered pairs used to designate them.

b. Prove that multiplication is commutative and left-distributive over addition in F.

c. Prove the statements made about negatives, unity, and inverses in the proof. Characterize all the elements of the class $E(e,e)$.

d. Prove that the mapping $\alpha : a \to E(a,e)$ of D onto D' is an isomorphism.

4-65. Show that $E(a,b) = E(ca,cb)$ if $c \neq 0$, and use this to show that a commutative ring without zero-divisors can be embedded in a field. (In other words, the unity element of D is not necessary to the proof of Theorem 4-22.)

4-66. Show that the mapping β defined in the paragraph immediately preceding these exercises is an isomorphism which preserves order.

4-67. Show that the construction of Theorem 4-22 yields nothing new if D is already a field. (Show that the field of quotients of a field is isomorphic to the original field.)

Miscellaneous exercises 4-68 to 4-76

4-68 to 4-70. (***An embedding theorem***) Every ring can be embedded in a ring with unity.

4-68. Consider $A \times J$ with addition and multiplication defined by $(a,n) + (b,m) = (a + b,\ n + m)$ and

$$(a,n)(b,m) = (ab + ma + nb,\ nm)$$

Show that $\langle A \times J, +, \ \rangle$ is a ring.

4-69. Show that $(0,1)$ is a unity element in $A \times J$.

4-70. Show that $\{(a,0) \mid a \in A\}$ is a subring of $A \times J$ isomorphic to A.

4-71 to 4-76. (***Approximate distributivity***)† A system $\langle A, +, \ \rangle$ is called a *w-ring* iff $\langle A, + \rangle$ is an abelian group, multiplication is associative in A, and the *approximate distributive laws* hold:

> There exist fixed elements d_1 and d_2 of A such that $x(y + z) = xy + xz - d_1$ and $(y + z)x = yx + zx - d_2$ for all $x,y,z \in A$.

☆*4-71.* In a w-ring $d_1 = 00 = d_2$. (Call this common element d.)

4-72. For all $x \in A$, $x0 = d = 0x$.

☆*4-73.* For all $x \in A$, $xd = d = dx$.

4-74. For all $x,y,z \in A$, $x(y - z) = xy - xz + d$ and

$$(y - z)x = yx - zx + d$$

4-75. The set $S = \{x \mid xy = yx = d \text{ for all } y \in S\}$ is a subgroup of the additive group of A, and $d \in S$.

4-76. Define $x \circ y = xy - d$ for all $x,y \in A$. Then $\langle A, +, \circ \rangle$ is a ring.

† These results are contained in a paper by Tôru Saitô, "Note on the Distributive Law," *American Mathematical Monthly*, p. 280, April, 1959. See also problem 4784 on page 318 of that issue.

5 POLYNOMIALS

The student of high school algebra spends a large part of his time working with polynomials—dividing them, factoring them, graphing them, solving equations in them. A little reflection on these chores should convince the more advanced student that the success or failure and difficulty of such efforts depend largely on the nature of the coefficients. For example, it is impossible to factor $x^2 + 25$ as a product of first-degree polynomials if only integer (or even real) coefficients are permitted; but it can be factored as

$$x^2 + 25 = (x + 5i)(x - 5i)$$

if complex coefficients are allowed.

Perhaps it seems artificial that in most beginning work complex coefficients are not allowed in factoring problems but complex roots of polynomial equations are as eagerly sought as real roots. There seems to be a sort of double standard at work. In this chapter, we shall attempt to study polynomials in such a way as to emphasize the role of the set of possible coefficients. (It is possible to do this so that a polynomial effectively *is* a sequence of coefficients. We shall not attempt this abstract "bare bones" approach at first but shall later sketch it in a collection of exercises.) Additionally, we shall develop most of the theory in the context of systems closely related to the number systems. We shall make the basic definitions for an arbitrary commutative ring, but the most important theorems will be stated and proved for integral domains and fields. Many of these theorems will be true in weaker systems.

Actually, problems about polynomials fall into two classes. The first is the manipulative class of problems—addition, division, factoring, etc. The second is the evaluative class—solving equations and things of that sort. The two types of problems are, of course, not unrelated, as anyone who has solved a polynomial equation by factoring can testify.

5-A COEFFICIENTS

The introductory definitions are motivated by the examples of polynomials with integer coefficients, such as $x^2 - 5x + 2$ which, for $x = 3i$ or $x = \frac{1}{2}$, takes on values which are decidedly not integral. Thus we consider the system of permissible coefficients as a subsystem of a larger system which includes "x."

Definition 5-1

Let $\langle B,+, \ \rangle$ be a commutative ring. Let A be a subring of B and let $x \in B$. A **polynomial** in x with *coefficients* in A (a polynomial in x *over* A) is an element of B of the form

$$f = a_0 + \sum_{i=1}^{n} a_i x^i = a_0 + a_1 x + a_2 x^2 + \cdots + a_n x^n$$

such that $a_0, a_1, \ldots, a_n \in A$ and $n \in J_0$. The largest integer k such that $a_k \neq 0$ is called the **degree** of f—denoted as "deg (f)"—and a_k is called the **leading coefficient** of f. (If $a_i = 0$ for all $i = 0,1, \ldots,$ then f has no degree.) Let $A[x]$ denote the set of all polynomials in x over A.

It should be noted that since $A[x] \subset B$, there is a notion of addition and multiplication of polynomials built into Definition 5-1.

Theorem 5-1

Let $f,g \in A[x]$ with

$$f = a_0 + \sum_{i=1}^{n} a_i x^i \qquad \text{and} \qquad g = b_0 + \sum_{i=1}^{m} b_i x^i$$

Then

1. $f + g = (a_0 + b_0) + \sum_{i=1}^{k} (a_i + b_i)x^i$, where $k = \max\{m,n\}$ and $a_j = 0$ if $j > n$ and $b_j = 0$ if $j > m$.

2. $fg = c_0 + \sum_{i=1}^{m+n} c_i x^i$, where $c_i = \sum_{r=0}^{i} a_r b_{i-r}$.

3. $\deg(f + g) \leq \max\{\deg(f), \deg(g)\}$ if these numbers exist.

4. $\deg(fg) \leq \deg(f) + \deg(g)$ if these numbers exist.

The proof of this theorem is deferred to the exercises, where the reader is also invited to explain why "$<$" can hold in parts 3 and 4. Part 2 of Theorem 5-1 would hold even if B were not a commutative ring, in which case the order of the a's and b's in the formula for fg would be significant—and it would be necessary to assume that x commutes with all elements of A. An important thing to notice about the formula for c_i is that the sum of the indices is always i.

Theorem 5-2

Let A be a subring of the commutative ring B. If $x \in B$, then $\langle A[x],+,\ \rangle$ is a ring. If B contains a unity element e and if $e \in A$, then $A[x]$ is the smallest subring of B containing both A and the element x. The ring A is a subring of $A[x]$.

PROOF. Beginning with the second conclusion: Every subring of

B which contains x and A must contain all powers x^k of x ($k \in J_+$) and all multiples $a_k x^k$ of these by elements of A. Further, it must contain all finite sums of such elements and elements of A. Thus it must contain all elements of the form

$$a_0 + a_1 x + \cdot \cdot \cdot + a_n x^n \qquad a_0, a_1, \ . \ . \ . \ , a_n \in A$$

so $A[x]$ is a subset of every subring of B which contains A and x.

But $A[x]$ itself is a ring. The closure properties follow from

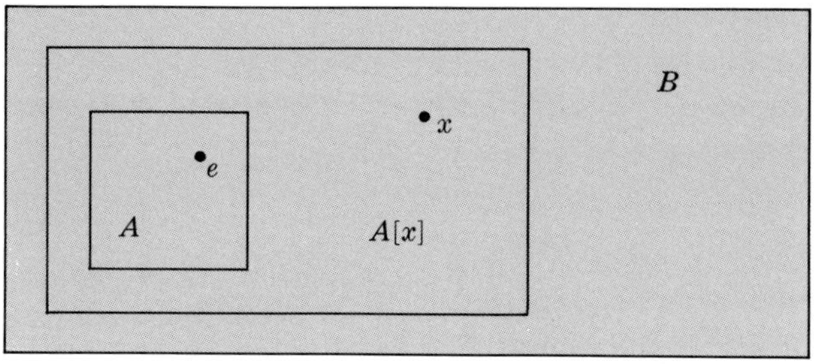

Figure 5-1 Structure of $A[x]$ ***if*** $e \in A$.

Theorem 5-1; necessary commutative, associative, and distributive properties are inherited from B; also $0 = 0 + \sum_{i=1}^{n} 0x^i$ and

$$-f = -a_0 + \sum_{i=1}^{n} (-a_i)x^i$$

are in $A[x]$.

If $a \in A$, then $a = a + \sum_{i=1}^{n} 0x^i$, and $a \in A[x]$. Thus A is a subset (and a subring) of $A[x]$. (See Figure 5-1.) ||

However, if $x \in A$, then $A[x] = A$ (a most uninteresting development, but one which has some significance for parts of the theory of polynomials). In general, if $e \notin A$, then $ex = x$ is not necessarily an element of $A[x]$.

At this point it is desirable to make a distinction between two types of x's which can be used to create $A[x]$. As an example to show the difference, consider $J[\sqrt{2}]$ and $J[\pi]$. In $J[\sqrt{2}]$, $1(\sqrt{2})^2 - 2 = 0$ and $(\sqrt{2})^3 - 2\sqrt{2} + 5 = (\sqrt{2})^2 + 3$, but in $J[\pi]$ the only polynomial equal to 0 is the polynomial $0 + \sum_{i=1}^{n} 0\pi^i$. (This last fact is very difficult to prove and was not known at all until 1882.)

Definition 5-2

Let A be a subring of a commutative ring B. An element $x \in B$ is **transcendental** over A iff the equation

$$a_0 + a_1x + \cdots + a_nx^n = 0$$

is true only if $a_0 = a_1 = \cdots = a_n = 0$. An element $y \in B$ is **algebraic** over A iff there exists an equation

$$a_0 + a_1y + \cdots + a_ny^n = 0$$

with at least one a_i not zero. (That is, y is algebraic over A iff it is not transcendental over A.)

This distinction between algebraic and transcendental elements will help, as we shall see in Theorem 5-3, because in dealing with transcendental elements we have unique representations for elements of B—a very desirable quality in doing polynomial arithmetic.

Note that every $b \in A$ is algebraic over A since, with $a_0 = b^2$ and $a_1 = -b$,

$$b^2 + (-b)b^1 = 0$$

[If $e \in A$, $b + (-e)b^1 = 0$ is a simpler example.] Referring to the earlier examples, $\sqrt{2}$ is algebraic over J and π is transcendental over J. However, π is algebraic over R, so this concept depends strongly on the coefficient set.

Theorem 5-3

Let A be a subring of a commutative ring B. An element x is transcendental over A iff the expression of each polynomial in x over A has unique coefficients.

PROOF. We shall prove first that if x is transcendental over A, two polynomials in x represent different elements of B. To this end, let

$$a_0 + \sum_{i=1}^{n} a_ix^i = b_0 + \sum_{i=1}^{n} b_ix^i$$

with n an integer at least as large as the greater of the two degrees. Then

$$(a_0 - b_0) + \sum_{i=1}^{n} (a_i - b_i)x^i = 0$$

Since x is transcendental over A, $a_i - b_i = 0$ for $i = 0,1, \ldots, n$. Thus $a_i = b_i$ for all i, and the polynomial representations are identical.

Conversely, if all polynomial representations are unique, then the polynomial representation of 0 is unique, so

$$a_0 + \sum_{i=1}^{n} a_i x^i = 0 = 0 + \sum_{i=1}^{n} 0x^i$$

implies that $a_0 = a_1 = \cdots = a_n = 0$. Thus x is transcendental over A. ||

This theorem tells us that the study of polynomials in transcendental elements is somewhat the simpler of the two cases since polynomials in x with different coefficients actually are different elements of the ring B. We have already observed that $A[x]$ is an extension of A in the sense of Definition 4-12 (see Theorem 5-2). The fact that there is only one element x annexed to A to generate the ring $A[x]$ leads us to call $A[x]$ a *simple transcendental extension* of A.

Theorem 5-4

Let x and y be transcendental over the ring A. Then $A[x]$ and $A[y]$ are isomorphic rings.

This theorem, whose uncomplicated proof is deferred to the exercises, tells us that there is essentially only one simple transcendental extension of A.

Exercises 5-1 to 5-11

5-1. Find the sum and product of the elements of $J[x]$, $4 - 7x^2 + 8x^3 - 2x^4$ and $3 + 2x - 3x^2 - 1x^3$.

5-2. Prove Theorem 5-1.

☆*5-3.* Give examples to show that "<" can hold in formulas 3 and 4 of Theorem 5-1.

5-4. Let f and g be given as in Theorem 5-1 and let $h = d_0 + \sum_{i=1}^{p} d_i x^i$. Characterize the coefficient of x^k in fgh.

5-5. Show that A is not a proper ideal in $A[x]$, and $A[x]$ is not a proper ideal in B. (A *proper* ideal in B is an ideal which is not B or $\{0\}$.)

5-6. Let $a \in A$, $a \neq 0$. Let x be transcendental over A. Under what circumstances is ax transcendental over A?

5-7. Let x be transcendental over A. Then $A[x]$ is a ring with unity iff A is a ring with unity.

5-8. Let x be transcendental over A. Then $A[x]$ is an integral domain iff A is an integral domain. If A is an integral domain and $f,g \in A[x]$, then $\deg(fg) = \deg(f) + \deg(g)$.

5-9. Prove Theorem 5-4.

5-10. Show that the following complex numbers are algebraic over J:
a. $\sqrt{3}\, i$ b. $\sqrt[3]{2}$ ☆c. $2 + \sqrt{5}$

5-11. Consider relationships between $A[x][y]$, $A[y][x]$, and $A[x,y]$, the smallest ring containing A, x, and y. Under what circumstances are two or more of these isomorphic?

5-B POLYNOMIALS AS SEQUENCES

This section outlines briefly, in exercise form, another approach to the introduction of the concept of a polynomial over A. The section may be omitted without danger, but its approach displays most vividly the role of the coefficients in the study of polynomials.

Definition 5-1 (Alternative form)

Let A be a ring. A **polynomial** over A is an infinite sequence of elements of A with only a finite number of nonzero terms. The elements of the sequence are called the **coefficients** of the polynomial.

Thus a polynomial is an expression of the form

$$(a_0,a_1,a_2, \ldots ,a_n,0,0,0, \ldots)$$

which can be identified with the expression $a_0 + \sum_{i=1}^{n} a_i x^i$, as considered earlier. Let A^* denote the set of polynomials over A and let

$$a(a_0, \ldots, a_n, 0, 0, \ldots) = (aa_0, \ldots, aa_n, 0, 0, \ldots)$$

for all $a \in A$ and all $(a_0, \ldots, a_n, 0, 0, \ldots) \in A^*$.

Exercises 5-12 to 5-18

5-12. Define equality, addition, multiplication, and degree in A^*. (Use Theorem 5-1 as a guide.)

5-13. Show that $\langle A^*,+,\ \rangle$ is an extension of A (in the sense of Definition 4-12).

5-14. Show that $\langle A^*,+,\ \rangle$ is a ring. (There is a rather obvious isomorphism with $A[x]$, for x transcendental over A, discernible at this point.)

5-15. Let A be a ring with unity e. Define $x = (0,e,0,0, \ldots)$. Show that $x^2 = (0,0,e,0,0, \ldots)$ and $x^n = (0,0, \ldots, 0,e,0, \ldots)$ with nth coefficient $[(n + 1)$st term$]$ e, for all $n \in J$.

5-16. Show that x commutes with every $a \in A$ and that

$$a_0(e,0,0, \ldots) + \sum_{i=1}^{n} a_i x^i = (a_0, a_1, \ldots, a_n, 0, 0, \ldots)$$

5-17. Show that x is transcendental over A (assuming that you defined equality to be identity in Exercise 5-12).

5-18. With x transcendental over J, is $P_1 = \left\{f = a_0 + \sum_{i=1}^{n} a_i x^i \mid a_0 > 0\right\}$ a positive set in $J[x]$? Is $P_2 = \{f \mid$ the leading coefficient of f is positive$\}$ a positive set in $J[x]$?

5-C POLYNOMIALS OVER INTEGRAL DOMAINS

We now investigate a division process as it relates to polynomials and consider such things as factoring and solving polynomial equations.

In doing this, we select as a basic system the integral domain so that the results will apply to all the usual number systems. We shall, throughout this section, let D denote an integral domain with at least two elements and assume that D is a subdomain of an integral domain B, which will serve the same purpose as the ring B of Section 5-A. From Exercise 5-8 we known that $D[x]$ is an integral domain if D is an integral domain.

All of the results in this section depend upon the existence and uniqueness of a division process for polynomials. It will be easier to understand the proof of the following theorem, the *division algorithm*, if the usual laborious division process for polynomials is kept in mind.

$$
\begin{array}{r|llll}
 & 2x & +\ 7/2 & & \\
\cline{2-5}
2x^2 - 3x + 6 & 4x^3 & +\ x^2 & -\ 4x & +\ 11 \\
 & 4x^3 & -\ 6x^2 & +\ 12x & \\
\cline{2-5}
 & & 7x^2 & -\ 16x & +\ 11 \\
 & & 7x^2 & -\ 21/2\,x & +\ 21 \\
\cline{3-5}
 & & & -\ 11/2\,x & -\ 10
\end{array}
$$

Definition 5-3

A polynomial $f \in D[x]$ is **monic** iff the leading coefficient of f is e, the unity element of D.

Since the division process and the proof of the next theorem are complicated, we shall prove the theorem first for the case in which the divisor is a monic polynomial and later establish a slightly more general result as a corollary.

Theorem 5-5 (Weak division algorithm)

Let D be an integral domain and let x be transcendental over D. Let $f = a_0 + \sum_{i=1}^{n} a_i x^i$ and $g = b_0 + \sum_{i=1}^{m} b_i x^i$ be elements of $D[x]$ such that $a_n \neq 0$ and $b_m = e$ in D. Then there exist unique polynomials q and r in $D[x]$ such that

$$f = qg + r$$

and

$$\deg (r) < \deg (g) = m$$

if deg (r) exists.

DISCUSSION. The existence of q and r (called the *quotient* and *remainder*) is established by reproducing the division process as commonly used, so that the diagram below will serve as a sketch of the existence proof.

$$
\begin{array}{r|l}
 & a_n x^{n-m} + (a_{n-1} - a_n b_{m-1}) x^{n-m-1} \\
\hline
x^m + b_{m-1}x^{m-1} + \cdots + b_0 & a_n x^n + a_{n-1}x^{n-1} + \cdots \qquad + a_0 \\
 & a_n x^n + a_n b_{m-1} x^{n-1} + \cdots + a_n b_0 x^{n-m} \\
\cline{2-2}
 & (a_{n-1} - a_n b_{m-1}) x^{n-1} + \cdots \\
 & (a_{n-1} - a_n b_{m-1}) x^{n-1} + \cdots \\
\cline{2-2}
 & \qquad \cdots \\
\cline{2-2}
 & \qquad r
\end{array}
$$

PROOF. If $n = \deg(f) < \deg(g) = m$, let $q = 0$ and $r = f$; thus $f = 0g + f$ is the required formula.

If $n \geq m$, then let

$$f' = f - (a_n x^{n-m})g$$

Now the nth coefficient of f' is $a_n - a_n = 0$, so $\deg(f') \leq n - 1$. If $\deg(f') < m$, let $q = a_n x^{n-m}$ and $r = f'$. But if $\deg(f') = k \geq m$ and the leading coefficient of f' is $c_k \neq 0$, repeat the division process (dividing $c_k x^k$ by x^m). Let $f'' = f' - (c_k x^{k-m})g$. If $\deg(f'') < m$, substitution into the displayed equation above yields

$$f'' + (c_k x^{k-m})g = f - (a_n x^{n-m})g$$

which, when solved for f, gives the required q and r.

If $\deg(f'') \geq m$, repeat the division process. Since the degree of the primed f's decreases at each step, the polynomials q and r will be obtained in a finite number of steps (no more than $n - m + 1$).

To show that the q and r obtained above are unique, assume $f = qg + r = q'g + r'$ with $\deg(r) < m$ and $\deg(r') < m$. Then $(q - q')g = r' - r$, where $\deg(r' - r) < m$. However,

$$\deg(r' - r) = \deg(q - q') + \deg(g) = \deg(q - q') + m \geq m$$

which is impossible if these degrees exist. Therefore, $\deg(r' - r)$ does not exist, so $r' - r = 0$, which implies in turn, since $D[x]$ is an integral domain and $g \neq 0$, that $q - q' = 0$. ||

As indicated above, this is not the strongest possible form of the theorem. Actually the division process can be carried out whenever the leading coefficient of g has an inverse. (See the sample division preceding Definition 5-3, where the existence of 2^{-1} was necessary to the completion of the process.)

Corollary 5-5*a* (Division algorithm)

Let D be an integral domain and let x be transcendental over D. Let $f = a_0 + \sum_{i=1}^{n} a_i x^i$ and $g = b_0 + \sum_{i=1}^{m} b_i x^i$ be elements of $D[x]$ such that $a_n \neq 0$ and b_m^{-1} exists in D. Then there exist unique polynomials q and r in $D[x]$ such that $f = qg + r$ and $\deg(r) < m$ if $\deg(r)$ exists.

PROOF. If b_m^{-1} exists, then $b_m^{-1}g$ is a monic polynomial satisfying the hypotheses of Theorem 5-5. Thus there exist p and r in $D[x]$ such that

$$f = p(b_m^{-1}g) + r$$

Thus $f = (b_m^{-1}p)g + r$.

The uniqueness of this quotient $b_m^{-1}p$ and remainder r is proved exactly as in Theorem 5-5. ||

Again we have avoided introducing a division operation into the integral domain $D[x]$. The result of the division algorithm has been stated as $qg + r$ using only the operations already defined in $D[x]$. Of course, since $D[x]$ is an integral domain, it can be extended to a field of quotients by the construction used in Theorem 4-22. The ordered pairs (f,g) of polynomials in x over D are sometimes called the *rational expressions* in x over D. This field of quotients is the smallest field "containing" D and x.

We now turn our attention to some important results involving "values" of polynomials. We have considered a polynomial previously as an element of a ring $A[x]$, but now our attitude is that of investigating the roles of different x's in the same polynomial form. In slightly different language, we have previously held x fixed and let the coefficients vary; now we propose to hold the coefficients fixed and let x vary. To do this efficiently and honestly, we introduce the concept of a polynomial function.

Definition 5-4

Let A be a commutative ring and let $f = a_0 + \sum_{i=1}^{n} a_i x^i$ be an element of $A[x]$. The **polynomial function** $\bar{f}$ corresponding to f is defined by the formula

$$b\bar{f} = a_0 + \sum_{i=1}^{n} a_i b^i$$

The domain of $\bar{f}$ may be taken to be the underlying ring B of Definition 5-1 or the ring A. In the second case, $\bar{f}$ is a mapping (function) from A into A. It may also be, but is not necessarily, onto A.

A note of caution is in order here. Different polynomials may define the same polynomial function on A. This is a relatively commonplace occurrence in finite rings. For example, $f = x^3 + x$ and $g = 2x$ in $J_3[x]$ determine the same function on $J_3 = \{0,1,2\}$ since $0\bar{f} = 0\bar{g} = 0$, $1\bar{f} = 1\bar{g} = 2$, and $2\bar{f} = 2\bar{g} = 1$ in J_3.

Theorem 5-6

Let f and g be two elements of $D[x]$ with D an integral domain, and let $\bar{f}$ and $\bar{g}$ be the corresponding polynomial functions with domain B an integral domain containing D. Then $a\bar{f} + a\bar{g} = a(\overline{f + g})$ and $(a\bar{f})(a\bar{g}) = a(\overline{fg})$ for all $a \in B$.

The proof of this theorem is a pure calculation to show that the expressions in the equations are equal elements of B. The theorem motivates the following definition—which, incidentally, makes a ring out of the set of all polynomial functions on D.

Definition 5-5

If $f,g \in D[x]$ with D an integral domain, define $\overline{f + g} = \bar{f} + \bar{g}$ and $\overline{fg} = \bar{f}\,\bar{g}$.

These definitions are consistent with the ones usually given for the sum and product of functions in the case in which an addition and multiplication are defined in the range of the functions (such as real-valued functions). In particular, with reference to the division algorithm, $\bar{f} = \overline{qg + r} = \bar{q}\,\bar{g} + \bar{r}$ in the sense that, for all $a \in B$, $a\bar{f} = (a\bar{q})(a\bar{g}) + (a\bar{r})$.

Exercises 5-19 to 5-24

5-19. Prove Theorem 5-6.

5-20. Find three distinct polynomials in $J_2[x]$, for each of which the corresponding polynomial function is equal to 0 everywhere on J_2.

5-21. Find two distinct polynomials in $J_5[x]$ which determine the same polynomial function.

5-22. Give examples of polynomials in $R[x]$ such that the corresponding polynomial functions do and do not map R onto R. Give examples of polynomial functions that are and are not one-to-one on R.

5-23. Find a polynomial function which is a permutation (not the identity) of J_5.

5-24. Find the q and r of Corollary 5-5*a* for $f = x^3 + 2x^2 + x + 1$ and $g = 2x + 1$ in

a. $R[x]$ ☆b. $J_3[x]$ c. $J_5[x]$

▲

Two classic results of importance in the theory of polynomial equations in $R[x]$ are the *remainder theorem* and its corollary, the *factor theorem,* which are true for systems less elaborate than R. The remainder theorem is an outgrowth of the following considerations.

In the division algorithm (Theorem 5-5) if g is a monic polynomial of degree 1 (in other words, $g = x - b$ for some $b \in D$), then the remainder r obtained by dividing f by g must be zero or have degree zero. In either case, $r \in D$. But in considering the value of the polynomial function $\bar{f}$ at $b \in D$, this value will also be an element of D. The remainder theorem asserts that this value is also r. Thus the division process with divisor $g = x - b$ and the polynomial function $\bar{f}$ both lead to the same element of D. For example, if $f = x^2 + 3x - 6$ in $J[x]$ and $g = x - 2$, then the division algorithm yields

$$x^2 + 3x - 6 = (x + 5)(x - 2) + 4$$

Also $2\bar{f} = 2^2 + 3 \cdot 2 - 6 = 4$.

Theorem 5-7 (Remainder theorem)

Let D be an integral domain and let f and $g = x - b$ be elements of $D[x]$. Then the remainder on dividing f by $x - b$ is $b\bar{f}$. [More precisely, if $f = qg + r$ with $r = 0$ or $\deg(r) < \deg(g) = 1$, then $r = b\bar{f}$.]

PROOF. Since $r = 0$ or $\deg(r) = 0$, r is a constant polynomial, an element of D; thus $d\bar{r} = r$ for all $d \in D$. Note that $b\bar{g} = b - b = 0$. Thus we obtain

$$b\bar{f} = b(\overline{qg + r}) = (b\bar{q})(b\bar{g}) + b\bar{r} = (b\bar{q})0 + b\bar{r} = r \quad \|$$

This theorem has applications to finding the values of polynomial functions for polynomials of large degree, where direct evaluation is tedious. This division process can be accelerated in $R[x]$ by a process known as *synthetic division,* which is discussed in most books titled "College Algebra."

An important corollary to the remainder theorem is the factor theorem, which is the basis of solving polynomial equations by factoring. The language of solutions of polynomial equations leads us to say that b is a *root* of the equation $x\bar{f} = 0$ iff $b\bar{f} = 0$ is true.

Corollary 5-7*a* (Factor theorem)

Let D be an integral domain and let f and $g = x - b$ be elements of $D[x]$. Then $g \mid f$ if, and only if, $b\bar{f} = 0$.

PROOF. By the division algorithm there exist $q,r \in D[x]$ such that $f = q(x - b) + r$. If $b\bar{f} = 0$, then

$$0 = b\bar{f} = (b\bar{q})(b - b) + r = 0 + r$$

Thus $r = 0$, which implies that $f = q(x - b)$; so $(x - b) \mid f$.

Conversely, if $(x - b) \mid f$, then there exists $h \in D[x]$ such that $f = h(x - b)$. But then

$$b\bar{f} = (b\bar{h})(b - b) = (b\bar{h})0 = 0 \quad \|$$

Exercises 5-25 to 5-32

5-25 to 5-27. ***(Some consequences of the factor theorem)***

5-25. Let D be an integral domain with $d_1, d_2, \ldots, d_n \in D$. Let

$f \in D[x]$. Then $d_1, d_2, \ldots, d_n$ are distinct roots of $x\bar{f} = 0$ iff $(x - d_1)(x - d_2) \cdots (x - d_n)$ divides f.

5-26. If $f \in D[x]$ and $\deg(f) = n$, then $x\bar{f} = 0$ has at most n distinct solutions in D.

5-27. If $f,g \in D[x]$ such that $n = \deg(f) \geq m = \deg(g)$, and if $d_i\bar{f} = d_i\bar{g}$ for $n + 1$ distinct elements $d_i \in D$, then $f = g$.

5-28. Verify the remainder theorem for the case

$$f = x^5 - 5x^3 + 2x^2 - x + 3$$

and $g = x - 2$ in $J[x]$.

5-29. Does Corollary 5-5*a* (the division algorithm) hold in $J_6[x]$? If so, do the remainder and factor theorems hold there? What is the status of these theorems in commutative rings that are not integral domains?

5-30. Is the process of synthetic division valid in $J_5[x]$? In every $J_n[x]$? In every commutative ring?

5-31. Is $x - 1$ a factor of $x^4 + 2x^3 + x^2 + x + 1$ in

a. $J_3[x]$ b. $J_5[x]$ c. $J[x]$

☆*5-32.* Prove Theorem 1-6. (Prove first for $n > 0$ and $d > 0$.)

5-D POLYNOMIALS OVER FIELDS

In this section F will always denote a field and we shall usually take x to be a transcendental element over F to assure uniqueness of the polynomials. In considering polynomials over a field F we can use all the previous results since all those hypotheses are satisfied by a field. If F is a field, $F[x]$ is not a field in general (see Theorem 5-8 below) but it is an integral domain by Exercise 5-8 and Theorem 4-20. Our principal topic here will be divisibility of polynomials. By Theorem 4-9 the units of an integral domain are precisely those elements that have inverses.

Theorem 5-8

Let F be a field and x transcendental over F. The polynomial $f \in F[x]$ is a unit of $F[x]$ iff f is a nonzero element of F (a constant polynomial).

PROOF. If $f \in F \subset F[x]$ and $f \neq 0$, then $f^{-1} \in F \subset F[x]$, so f is a unit. Conversely, if f is a unit of $F[x]$, then there exists $g \in F[x]$ such that $fg = e$, the unity of $F[x]$ (and of F). But

$$0 = \deg (e) = \deg (f) + \deg (g)$$

so $\deg (f) = 0$. Thus $f \in F$ and $f \neq 0$. ||

We have thus proved that $F[x]$ is not a field since no polynomial of degree 1 or greater has a multiplicative inverse. We also have established that there are integral domains which contain more units than just e and $-e$ but are not fields.

In the integers there is the familiar concept of a greatest common divisor of two integers. We shall try below to define a corresponding concept for polynomials and develop some of the related theory. There are a large number of topics in the theory of polynomials over fields which we shall not have space to mention. The interested reader is invited to refer to any of the several more encyclopedic texts in this area for more information.

Actually we can, and will, simplify our discussion somewhat by attending mainly to polynomials with leading coefficient e (called *monic* polynomials in Definition 5-3). This will not disturb the generality of our results since every nonzero polynomial has an associate (see Definition 4-8) which is monic. That is, if the leading coefficient of $f = a_0 + \sum_{i=1}^{n} a_i x^i$ is a_n, then $g = a_n{}^{-1}f$ has leading coefficient e and $g \mid f$ and $f \mid g$. (Compare part 4 of Theorem 4-9.)

Definition 5-6

Let $f,g \in F[x]$ with at least one of f and g not zero. Then the polynomial $d \in F[x]$ is called a **greatest common divisor** (gcd) of f and g iff

1. d is a monic polynomial.
2. $d \mid f$ and $d \mid g$.
3. If $h \in F[x]$ such that $h \mid f$ and $h \mid g$, then $h \mid d$.

The next theorem is the correspondent for $F[x]$ of Theorem 4-17 for J. The proof of Theorem 4-17 is similar to that given below with the gcd (a number) itself replacing the degree of the gcd of the polynomials.

Theorem 5-9

Let x be transcendental over a field F. Any two polynomials f and g, not both zero, in $F[x]$ have a unique gcd which is the monic nonzero polynomial d of minimum degree of the form

$$d = sf + tg \qquad s,t \in F[x]$$

PROOF. Let $G = \{sf + tg \mid s,t \in F[x]\}$. Either f or g is not zero. If $f \neq 0$, let $s = e$ and $t = 0$, so that $sf + tg = f$. Thus G contains nonzero polynomials. If $sf + tg \in G$ and $sf + tg \neq 0$ with leading coefficient a, then $(a^{-1}s)f + (a^{-1}t)g$ is a monic element of G. Thus G contains monic polynomials. Let d be a monic polynomial of minimum degree in G.

We now show that the d described above is unique. If d and d_1 are monic polynomials of the same minimum degree in G, then $d = sf + tg$ and $d_1 = s_1f + t_1g$. Assume $d - d_1 \neq 0$; then

$$d - d_1 = (s - s_1)f + (t - t_1)g$$

is in G and, since both d and d_1 have leading coefficient e, deg $(d - d_1) <$ deg (d), contradicting the choice of d. So $d - d_1 = 0$.

By the division algorithm, let $f = qd + r$ with $q,r \in F[x]$ and deg $(r) <$ deg (d) or $r = 0$. Now

$$r = f - qd = f - q(sf + tg) = (e - qs)f + (-qt)g$$

is in G, but deg $(r) <$ deg (d) contradicts the choice of d again. Thus $r = 0$, and $d \mid f$. Similarly, $d \mid g$.

If d' is any common divisor of f and g, then $f = hd'$ and $g = kd'$ and

$$d = sf + tg = s(hd') + t(kd') = (sh + tk)d'$$

Thus $d' \mid d$ and d is a gcd of f and g. The proof that there can be only one gcd of f and g, no matter how determined, is left as an exercise. ||

We can carry over to the present discussion the concepts of prime and composite as defined in Definition 4-9. Then $F[x]$ will also have the four categories of elements: zero, the units, primes, and composites. The terms **irreducible** and **reducible** are sometimes used in place of "prime" and "composite," respectively. Actually these new terms are usually defined in a slightly different way in expositions of weaker systems; here they are interchangeable with "prime" and "composite." We shall use "irreducible" and "reducible" in conformity with common practice.

A polynomial may be irreducible over one field and reducible over a larger field. For example, $x^2 - 2$ is irreducible in $Q[x]$ but reducible as $(x - \sqrt{2})(x + \sqrt{2})$ in $R[x]$. Since the units in $F[x]$ are the nonzero elements of F, it follows that the proper factors of a reducible (composite) polynomial are polynomials of degree 1 or greater. Thus $f \in F[x]$ is reducible iff there exist $s,t \in F[x]$ such that $\deg(s) \geq 1$, $\deg(t) \geq 1$, and $f = st$. In particular, since

$$\deg(f) = \deg(s) + \deg(t)$$

a polynomial of degree 1 is irreducible.

Exercises 5-33 to 5-38

5-33. Complete the proof of Theorem 5-9.

5-34. Find all irreducible polynomials of degree less than 3 in $J_5[x]$.

5-35. Look up the Euclidean algorithm for finding the gcd of two integers. Describe an analog for finding the gcd of two polynomials in $R[x]$.

5-36. Prove Theorem 4-17.

5-37. Use Theorem 4-17 to show that the only ideals in J are the principal ideals P_n (Exercise 4-23).

5-38. Show that x is transcendental over Q iff x is transcendental over J.

▲

We are now well on the way to a unique factorization theorem for polynomials similar to the theorem stated as Theorem 4-19. The sequence of theorems leading up to it will involve the analog of Theorem 4-18.

Theorem 5-10

If $p,f,g \in F[x]$ with x transcendental over F, and if p is irreducible and $p \mid fg$, then $p \mid f$ or $p \mid g$.

DISCUSSION. We have here a theorem with an "either . . . or" conclusion. We shall prove it by showing that if the hypothesis is satisfied but one part of the conclusion fails, then the other part of the conclusion holds.

PROOF. Assume $p \mid fg$ and $p \nmid f$. Then the gcd of p and f must be a monic divisor of p—namely, e or a monic associate of p—but cannot be an associate of p since $p \nmid f$. Thus gcd $\{p,f\} = e$ and there exist, by Theorem 5-9, $s,t \in F[x]$ such that $e = sp + tf$. Since $p \mid fg$, there exists $h \in F[x]$ such that $ph = fg$. Now

$$g = eg = (sp + tf)g = spg + t(fg) = spg + t(ph) = (sg + th)p$$

so $p \mid g$. ||

Corollary 5-10*a*

If p is an irreducible element of $F[x]$, and if $p \mid f_1f_2 \cdots f_n$, then $p \mid f_i$ for some i.

The inductive proof of this generalization of Theorem 5-10 is an exercise.

Now that we know that an irreducible polynomial must divide one factor if it divides a product, we can prove unique factorization.

Theorem 5-11 (Unique factorization theorem)

Let x be transcendental over a field F. Each polynomial $g \neq 0$ in $F[x]$ can be written as

$$g = a_m p_1 p_2 \cdots p_r$$

where a_m is the leading coefficient of g and each p_i is a monic irreducible polynomial. The factorization is unique except for the order of the factors.

PROOF. If deg $(g) = 0$, or if deg $(g) > 0$ and g is irreducible in $F[x]$, the result is obvious. So we proceed inductively on the degree of g. If deg $(g) = 0$ or 1, we are through. Assume that all polynomials of degree less than n can be factored as above, and let deg $(g) = n$. If g is irreducible, we are finished. If g is reducible, let $g = hk$ where $1 \leq$ deg $(h) < n$ and $1 \leq$ deg $(k) < n$. But then h and k can be factored with irreducible factors, according to the induction hypothesis, as

$$h = ap_1p_2 \cdots p_m \qquad \text{and} \qquad k = bq_1q_2 \cdots q_r$$

Thus

$$g = hk = (ab)p_1p_2 \cdots p_mq_1 \cdots q_r$$

is the required factorization.

To show that the factorization is unique, we will need Corollary 5-10*a*. Let

$$g = a_mp_1p_2 \cdots p_r = a_mq_1q_2 \cdots q_s$$

where the p's and q's are monic irreducible polynomials. Then $p_1 \mid q_1q_2 \cdots q_s$, and p_1 must divide one of the q's by Corollary 5-10*a*. Assume the order of q's such that $p_1 \mid q_1$. But q_1 is also both irreducible and monic, so $p_1 = q_1$. Cancel $p_1 = q_1$ and repeat the argument with p_2. In this way we see that $r = s$ and $p_i = q_i$ for all i. ||

Exercises 5-39 to 5-43

5-39. Prove Corollary 5-10*a*.

5-40. a. Prove Theorem 4-18.
b. Prove Corollary 4-18*a*.
c. Prove Theorem 4-19.

5-41. Factor into a product of irreducible factors in $Q[x]$, $R[x]$, and $C[x]$ (C denotes the complex numbers).
a. $x^3 + 3x^2 - x - 3$
b. $x^3 - 5x^2 + 5x - 1$
c. $x^4 - 25$
d. $x^6 - x^4 - x^2 + 1$

5-42. Factor the polynomial in 5-41*d* above into irreducible factors in $J_3[x]$.

5-43. The fundamental theorem of algebra (so-called) states that every polynomial equation $x\bar{f} = 0$, where f has degree 1 or greater in $C[x]$, has at least one complex number root. Show that this implies
a. Every polynomial of degree $n > 1$ in $C[x]$ can be expressed as a product of factors of degree 1 in $C[x]$.
b. If $f \in R[x] \subset C[x]$, deg $(f) = 2$, and $x\bar{f} = 0$ has only imaginary roots ($a + bi$ with $b \neq 0$), then f is irreducible in $R[x]$.

▲

In an attempt to make our study of polynomials well rounded we return to the elements algebraic over F and consider some properties of these algebraic elements. We shall use the polynomial function $\bar{f}$ corresponding to $f \in F[x]$ as a mapping with domain including all of the containing system B.

Theorem 5-12

Let F be a subfield of a field B and let $x,y \in B$. Assume x is transcendental over F and y is algebraic over F. Then there exists a unique monic irreducible polynomial $p \in F[x]$ such that $y\bar{p} = 0$. Also p divides every polynomial $f \in F[x]$ such that $y\bar{f} = 0$.

PROOF. Since y is algebraic over F, there exists $g \in F[x]$ such that $y\bar{g} = 0$ and $\deg(g) > 0$. (Compare Definition 5-2.) Let

$$H = \{g \in F[x] \mid \deg(g) > 0 \text{ and } y\bar{g} = 0\}$$

Then H contains monic polynomials (as in the proof of Theorem 5-9), and we may let p be a monic polynomial of minimum degree in H.

If p were reducible, we would have $p = hk$ with $1 \leq \deg(h) < \deg(p)$ and $1 \leq \deg(k) < \deg(p)$, which would yield

$$0 = y\bar{p} = y(\overline{hk}) = (y\bar{h})(y\bar{k})$$

so that $y\bar{h} = 0$ or $y\bar{k} = 0$. But this contradicts the choice of p as a polynomial in H of minimum degree. Thus p is irreducible.

If $g \in H$, then $g = qp + r$ with $\deg(r) < \deg(p)$ or $r = 0$. But then

$$y\bar{g} = 0 = (y\bar{q})(y\bar{p}) + y\bar{r} = (y\bar{q})0 + y\bar{r} = y\bar{r}$$

making $r \in H$ of degree less than $\deg(p)$ unless $r = 0$. So $r = 0$, and $p \mid g$ for all $g \in H$.

This result also shows that p is unique, for if p and p' are monic of minimum degree in H, then $p \mid p'$ and $p' \mid p$, so $p = p'$ since both are monic. ||

Definition 5-7

Let y be algebraic over a field F. The polynomial p described in Theorem 5-12 is called the **minimum polynomial** of y (with respect to F). The degree of p is the **degree** of y over F.

For example, if $F = Q$ and $y = \sqrt{2}$, $\sqrt{2}$ is a solution of all the equations $x^4 - 4 = 0$, $x^3 - \frac{1}{2}x^2 - 2x + 1 = 0$, $x^2 - 2 = 0$, etc.; but $x^2 - 2$ is the only monic polynomial of degree 2 or less for which this is true, and $x^2 - 2$ divides every polynomial f with rational coefficients such that $\sqrt{2}\,\bar{f} = 0$. Thus $x^2 - 2$ is the minimum polynomial of $\sqrt{2}$ over Q and $\sqrt{2}$ has degree 2 over Q. (However, $\sqrt{2}$ has degree 1 over R since $x - \sqrt{2}$ is its minimum polynomial there.)

Exercises 5-44 to 5-46

5-44. What is the degree of the complex number $a + bi$ over R?

5-45. Find the degree of each of the following over Q:

a. $\sqrt[3]{4}$ b. $\sqrt{2} + \sqrt{3}$

c. $\sqrt{2} + \sqrt[3]{3}$ d. $1 + i\sqrt[3]{2}$

(Compare Exercise 5-10.)

5-46. If y is algebraic of degree m over F, show that, for every polynomial $f \in F[x]$ (x transcendental), there exists a monic polynomial $g \in F[x]$ such that $\deg(g) < m$ and $y\bar{f} = y\bar{g}$. Thus the polynomials in an algebraic element y over F are all "essentially" of degree less than the degree of y since each, as an element of the containing field B, can be expressed as a polynomial of degree less than that of y.

5-E SOME EXAMPLES OF FIELDS

We already have seen some examples of fields: the real, rational, and complex number systems. In addition we have studied some finite fields, namely, J_p for p a prime positive integer. In this section we wish to exhibit a large class of other fields. One of our goals will be a formal development of the complex numbers, to which we have paid scant attention so far. To do these things, we shall first expand our general knowledge of properties of rings and fields. Our immediate goal is Corollary 5-16*a*.

Theorem 5-13

Let A be a ring. Let $\{I_t\}_{t\in K}$ be a collection of ideals in A. Then $\bigcap_{t\in K} I_t$ is an ideal in A.

The proof of this theorem is elementary, following from Theorem 4-3 in the same way that Exercise 2-18 follows from Theorem 2-6.

Definition 5-8

Let S be a subset of a ring A. The **ideal generated** by S is the set $[S] = \bigcap_{S \subset I} I$ such that I is an ideal containing S. (That is, $[S]$ is the smallest ideal in A containing S.)

The concept of $[S]$ is similar to that of the cyclic group $[a]$ as the smallest subgroup of a group containing the element a, or to $A[x]$ as the smallest ring containing A and x—and for this reason we use similar notation. If S is a one-element set $S = \{r\}$ we shall write $[r]$ instead of $[\{r\}]$. Actually the theory of the ideal $[r]$ is that of principal ideals if A is a commutative ring. In the notation of Exercise 4-23, $[r] = P_r$.

To develop an idea which will be helpful in the proof of the theorem following this one, we consider what happens if the set S consists of an ideal and one more element.

Theorem 5-14

Let A be a commutative ring with unity e. Let $r \in A$ and I an ideal in A. Then

$$[I \cup \{r\}] = \{ar + b \mid a \in A \text{ and } b \in I\}$$

PROOF. Let $C = \{ar + b \mid a \in A \text{ and } b \in I\}$. Let K be an ideal containing $I \cup \{r\}$. Then $ar \in K$ for all $a \in A$, and $b \in I \subset K$; thus $ar + b \in K$. So $ar + b \in \bigcap K = [I \cup \{r\}]$. This shows that $C \subset [I \cup \{r\}]$.

In order to show that $[I \cup \{r\}] \subset C$, we shall prove that C is an ideal containing $I \cup \{r\}$. First, $r = er + 0 \in C$, and for all $b \in I$, $b = 0r + b \in C$, so $I \cup \{r\} \subset C$. The set C is an ideal by Theorem 4-3 since $ar + b \in C$ implies that

$$-(ar + b) = (-a)r + (-b)$$

is in C, and that for all $d \in A$,

$$d(ar + b) = (da)r + (db)$$

is in C since db is in the ideal I and $da \in A$. ||

Definition 5-9

An ideal I in a ring A is a **maximal ideal** in A iff, for all ideals K such that $I \subset K \subset A$, either $K = I$ or $K = A$. (In other words, I is maximal iff the only ideal that properly contains I is A itself.)

For example, the ideal $[6] = \{6n \mid n \in J\}$ is not maximal in J since it is contained in [3]. However, [3] is a maximal ideal in J. In fact, the maximal ideals in J are precisely those of the form $[p]$ where p is a prime. It is the maximal ideals which can be used to construct examples of fields.

Theorem 5-15

Let A be a commutative ring with unity e, and let I be a maximal ideal in A. Then the quotient ring A/I is a field.

PROOF. By Theorem 4-6, A/I is a homomorphic image of A, which makes A/I a commutative ring with unity $e + I$ and zero $0 + I = I$. Let $r \notin I$; we shall show that $r + I$ has an inverse for multiplication in A/I.

Since I is itself maximal in A, $[I \cup \{r\}] = A$, so every element of A has the form $ar + b$ with $a \in A$ and $b \in I$ by Theorem 5-14. In particular, there exist $u \in A$ and $v \in I$ such that $e = ur + v$. Thus

$$(u + I)(r + I) = ur + I = (e - v) + I = e + I$$

since $v \in I$. Thus $(r + I)^{-1} = u + I$ and A/I is a field. ||

To apply this result, we first isolate some maximal ideals in $F[x]$.

Theorem 5-16

Let F be a field and p an irreducible polynomial in $F[x]$. Then $[p]$ is a maximal ideal in $F[x]$.

PROOF. The ideal $[p]$ is the set $\{fp \mid f \in F[x]\}$ of all multiples of p by elements of $F[x]$. Let K be an ideal in $F[x]$ such that $[p] \subset K \subset F[x]$. Assume $[p] \neq K$. Then there exists $g \in K$ such that $p \nmid g$. Since p is irreducible, gcd $\{p,g\} = e$ and there exist (by Theorem 5-9) polynomials s and t such that $e = ps + gt$. Since p and g belong to the ideal K, so do ps and gt; consequently $ps + gt = e$. Since $e \in K$, all multiples $fe = f$ of e belong to K for all $f \in F[x]$. Thus $K = F[x]$, and $[p]$ is maximal. ||

Corollary 5-16*a*

If p is irreducible in $F[x]$, then $F[x]/[p]$ is a field.

PROOF. Apply Theorem 5-16 to Theorem 5-15. ||

Exercises 5-47 to 5-51

5-47. Prove Theorem 5-13.

5-48. Prove that $[n]$ is a maximal ideal in J iff n is prime. (Compare the proof of Theorem 5-16.)

5-49 to 5-51. (***Prime ideals***) An ideal I in a ring A is a *prime ideal* iff, for all $a,b \in A$, if $ab \in I$, then $a \in I$ or $b \in I$.

5-49. Let A be a commutative ring with unity. Then every maximal ideal in A is a prime ideal in A.

5-50. Relate Exercise 5-49 to the ring of integers.

5-51. Show that $\{0\}$ is a prime ideal which is not maximal in J.

▲

If f is any polynomial in $F[x]$, by the division algorithm there exist $q,r \in F[x]$ such that $f = qp + r$. Since $qp \in [p]$, f is equivalent (modulo $[p]$) to its remainder r on division by p. Thus the elements of $F[x]/[p]$ are cosets in $F[x]$ of the form $r + [p]$ with $\deg(r) < \deg(p)$ or $r = 0$. It will be convenient to use the possible remainders r as representatives of their residue classes and reduce sums and products modulo the polynomial p—similar to our description of "clock arithmetic" of integers in Chapter 2.

As a first example consider the field $J_2[x]/[x^2 + x + 1]$. If $x^2 + x + 1$ is reducible in $J_2[x]$, then it must have a linear factor $x - d$ to which will correspond a root d of $x^2 + x + 1 = 0$. But neither $d = 0$ nor $d = 1$ is a root of $x^2 + x + 1 = 0$, so $x^2 + x + 1$ is irreducible in $J_2[x]$—a line of reasoning that works through $\deg(p) = 3$. The possible remainders are first-degree polynomials $ux + v$ where u and v are 0 or 1. Thus we can caricature $J_2[x]/[x^2 + x + 1]$ as a field of order 4 with elements $0 = 0x + 0$, $1 = 0x + 1$, $x = 1x + 0$, and $x + 1$ with addition and multiplication modulo $x^2 + x + 1$.

The tables for this field follow. Notice that the arithmetic on the coefficients is performed in J_2 where $1 + 1 = 0$ and $-1 = 1$.

$+$	0	1	x	$x+1$
0	0	1	x	$x+1$
1	1	0	$x+1$	x
x	x	$x+1$	0	1
$x+1$	$x+1$	x	1	0

	0	1	x	$x+1$
0	0	0	0	0
1	0	1	x	$x+1$
x	0	x	$x+1$	1
$x+1$	0	$x+1$	1	x

As an example, the product $x(x + 1) = 1$ is found as follows: (1) $x(x + 1) = x^2 + x$ in $J[x]$. (2) Divide $x^2 + x$ by $x^2 + x + 1$ giving $x^2 + x = 1 \cdot (x^2 + x + 1) - 1$. (3) Replace the remainder -1 by the equivalent 1 in J_2.

As a second example of a field, consider $J_3[x]/[x^2 + 1]$, which is a field with nine elements—remainders of the form

$$\_x + \_$$

where the blanks may be filled in by 0, 1, or 2. In this way, using Corollary 5-16*a*, one can construct a field whose order is a power n of a prime integer r by taking as the field of coefficients J_r and working with an irreducible polynomial $p \in J_r[x]$ such that $\deg(p) = n$. We omit discussions of the existence of the appropriate polynomial p.

An important and interesting example to be drawn from Corollary 5-16*a* is not a finite field. Let us begin with the real field R and consider $x^2 + 1 \in R[x]$. If $x^2 + 1$ is reducible in R, we may write

$$x^2 + 1 = (x - a)(x - b) = x^2 - (a + b)x + ab$$

Thus a and b must be real numbers whose sum $a + b = 0$ and whose product $ab = 1$. Since R is an ordered integral domain, this implies $b = -a$ and $a(-a) = -a^2 = 1$, an impossibility. Thus $x^2 + 1$ is irreducible in $R[x]$. Let us consider the structure of the field

$$R[x]/[x^2 + 1]$$

The typical elements are the remainders $a + bx$ of degree less than or equal to 1 with $a,b \in R$. Let us denote by K the field composed of these remainders, with sums and products taken modulo $x^2 + 1$. If $a + bx$ and $c + dx$ are in K, then

$$(a + bx) + (c + dx) = (a + c) + (b + d)x$$

To compute the product: $(a + bx)(c + dx) = ac + (ad + bc)x + bdx^2$ in $R[x]$; dividing this by $x^2 + 1$,

$$\begin{array}{r|l} & bd \\ \cline{2-2} x^2 + 1 & bdx^2 + (ad + bc)x + ac \\ & bdx^2 \qquad\qquad\qquad + bd \\ \cline{2-2} & \qquad (ad + bc)x + (ac - bd) \end{array}$$

so the product in K is

$$(a + bx)(c + dx) = (ac - bd) + (ad + bc)x$$

If we replace x by the imaginary unit i such that $i^2 = -1$, we discover that the formulas for addition and multiplication are just the usual complex number formulas. Thus K, or $R[x]/[x^2 + 1]$, is isomorphic to the complex numbers C—that is, K is algebraically the same as C.

Exercises 5-52 to 5-58

5-52. Why must $\deg(p) \leq 3$ in order for us to equate the statement that p is irreducible in $F[x]$ with the statement that $x\bar{p} = 0$ has no solutions in F?

5-53. Construct a field of order $2^3 = 8$. Give its addition and multiplication tables.

5-54. Show the existence of a field of order 27. Describe its elements.

5-55. Show the existence of fields of order 25, 125, and 49.

5-56. Does there exist a field of order 16?

5-57. Show that $Q[x]/[x^2 - 2]$ is a field. Describe its elements.

5-58. Discuss the system $R[x]/[x^2 + 3]$.

6 TOPICS IN GROUP THEORY

The purpose of this chapter is to extend some of the results of Chapters 2 and 3 by developing three topics in the theory of groups independent of applications to the number systems. It is hoped that the reader will see something of the flavor of abstract algebra in this development of ideas for their own sake. Although the ideas developed here do have applications in other areas of mathematics and are important to further developments in algebra, they are not as directly related to the number systems as were most of the ideas of Chapters 4 and 5.

We consider three topics—isomorphism, direct products, and conjugacy. A basic tool in this study is the fundamental theorem, 3-9.

Theorem 3-9

If α is a homomorphism of a group $\langle G,\circ\rangle$ with identity e onto a group $\langle \bar{G},*\rangle$ with identity $\bar{e}$ and if K is the kernel of α, then K is a normal subgroup of G and $\bar{G}$ is isomorphic to G/K. Conversely, if H is any normal subgroup of G, then G/H is a homomorphic image of G under the mapping $\beta : g \rightarrow g \circ H$.

In this connection it is worthwhile to note also Exercise 3-47: A group homomorphism α is an isomorphism iff its kernel $K = \{e\}$.

6-A ISOMORPHISM THEOREMS

The goal of this section is the proof of two theorems, 6-5 and 6-6, containing information about homomorphisms and normal subgroups similar to, but slightly more complicated than, Theorem 3-9. The initial results have some interest of their own but are viewed here as preliminary to the two main theorems.

Definition 6-1

If A and B are subsets of a group $\langle G, \ \rangle$, then

$$AB = \{ab \mid a \in A \text{ and } b \in B\}$$

We shall be concerned largely with the product set AB when A and B are both subgroups of G.

Theorem 6-1

Let H and K be subgroups of G with either H or K normal in G. Then HK is a subgroup of G.

PROOF. Assume that K is normal in G. We shall give a proof that HK is a subgroup by use of Theorem 2-6.

Since $H \neq \varnothing$ and $K \neq \varnothing$, it follows that $HK \neq \varnothing$.

Let $x = h_1k_1$ and $y = h_2k_2$ be elements of HK. Then

$$xy^{-1} = (h_1k_1)(h_2k_2)^{-1} = h_1k_1k_2^{-1}h_2^{-1}$$

Since K is normal in G, there exists an element $k \in K$ such that $(k_1k_2^{-1})h_2^{-1} = h_2^{-1}k$. Thus

$$xy^{-1} = h_1(h_2^{-1}k) = (h_1h_2^{-1})k$$

and $xy^{-1} \in HK$. ||

The proof of Theorem 6-1 for the case of H normal is a slight modification of that above.

Actually more can be said about HK if one of H and K is normal. In Exercise 2-18 we showed that the intersection of subgroups is a subgroup. No such theorem was presented for unions of subgroups for the simple reason that no such theorem is true. However, the union $H \cup K$ of two subgroups is contained in some smallest subgroup of G (some subgroup contained in every other subgroup containing $H \cup K$), and if one of H and K is normal, then that subgroup is HK.

Theorem 6-2

If H and K are subgroups of a group G and if H or K is normal in G, then HK is the smallest subgroup of G containing $H \cup K$.

Corollary 6-2*a*

Under the hypotheses of Theorem 6-2, $HK = KH$.

Exercises 6-1 to 6-5

6-1. Every subgroup of the abelian group $\langle J_6, +_6\rangle$ is normal. Calculate the product (that is, sum) of every pair of subgroups of J_6.

6-2. Find subgroups H and K in Table 1 of Appendix I such that HK is not a subgroup.

☆*6-3.* Show that the conclusion of Theorem 6-1 can hold on occasion when the hypothesis fails by exhibiting subgroups H and K of a group G such that neither H nor K is normal but HK is a proper normal subgroup of G.

6-4. Prove Theorem 6-2 and Corollary 6-2*a*.

6-5. Can AB be a subgroup of a group G if neither A nor B is a subgroup?

▲

To obtain the first main result of this section we need to recall Exercises 3-48 and 3-49. We state these results as a theorem.

Theorem 6-3

Let α be a homomorphism of a group G onto a group $\bar{G}$. If H is a normal subgroup of G, then

$$H\alpha = \{h\alpha \mid h \in H\}$$

is a normal subgroup of $\bar{G}$. If $\bar{H}$ is a normal subgroup of $\bar{G}$, then

$$\bar{H}\alpha^{-1} = \{h \in G \mid h\alpha \in \bar{H}\}$$

is a normal subgroup of G containing the kernel of α.

Thus normality is preserved in both directions by homomorphisms. (The notation $H\alpha$ and $\bar{H}\alpha^{-1}$ is consistent with Exercise 1-44.) Another result on normality needed immediately is the following.

Theorem 6-4

If H and K are subgroups of G with $K \subset H \subset G$ and if K is normal in G, then K is normal in H.

DISCUSSION. Note that a converse of this theorem is false (Exercise 3-35). The proof of the theorem is deferred to the exercises.

We are now ready to consider a more complicated result and a corollary. We shall use the notation

$$G \cong \bar{G}$$

to indicate that G is isomorphic to $\bar{G}$.

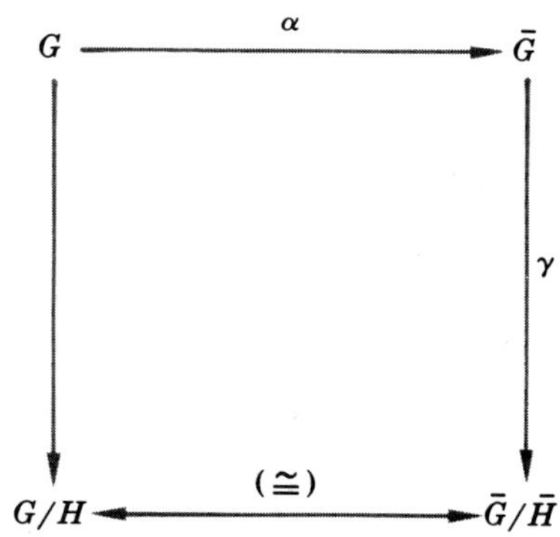

***Figure 6-1** Diagram of Theorem 6-5.*

Theorem 6-5

If α is a homomorphism of a group G onto $\bar{G}$ and if $\bar{H} = H\alpha$ is normal in $\bar{G}$ (or if H is normal in G and contains K) then

$$G/H \cong \bar{G}/\bar{H}$$

DISCUSSION. The diagram above (Figure 6-1) illustrates the relationships declared by the theorem.

PROOF. By Theorem 6-3, if H is normal in G, then $\bar{H}$ is normal in $\bar{G}$ (and conversely), so G/H and $\bar{G}/\bar{H}$ are both defined. By Theorem 3-9 the mapping $\gamma : \bar{G} \to \bar{G}/\bar{H}$ defined by $\bar{g}\gamma = \bar{g}\bar{H}$ is a homomorphism and $\bar{H} \subset \bar{G}$ is its kernel. Since α and γ are homomorphisms, the composite $\alpha\gamma$ is a homomorphism of G onto $\bar{G}/\bar{H}$. Now $h \in H$ iff $h\alpha \in \bar{H}$ iff

$$h(\alpha\gamma) = (h\alpha)\gamma = \bar{H}$$

(This is because γ maps elements of $\bar{H}$, and only elements of $\bar{H}$, onto the *element* $\bar{H}$ of $\bar{G}/\bar{H}$.) Thus H is the kernel of $\alpha\gamma$. By Theorem 3-9, since $\alpha\gamma$ maps G onto $\bar{G}/\bar{H}$ with kernel H, $G/H \cong \bar{G}/\bar{H}$. ||

Corollary 6-5*a*

Let K and H be subgroups of a group G with K and H normal in G and $K \subset H$. Then

$$G/H \cong (G/K)/(H/K)$$

PROOF. (This result is easily remembered as an intuitive sort of simplification of fractions.)

By Theorem 6-4 K is also normal in H. Thus G/H, G/K, and H/K are all defined. Let $\alpha : G \to \bar{G} = G/K$ be defined by $g\alpha = gK$. Then

$$\bar{H} = H\alpha = \{h\alpha \mid h \in H\} = \{hK \mid h \in H\} = H/K$$

Identification of α, $\bar{G}$, and $\bar{H}$ in Theorem 6-5 yields the conclusion

$$G/H \cong (G/K)/(H/K) \quad ||$$

The second isomorphism theorem is of a different type from the results above. It has to do with subgroups AB (the smallest subgroup containing $A \cup B$) and $A \cap B$ and a relationship between these.

Theorem 6-6

If A and B are subgroups of a group G and if A is normal in G, then $A \cap B$ is normal in B and

$$AB/A \cong B/(A \cap B)$$

PROOF. By Theorem 6-1 AB is a subgroup of G, and by Exercise 2-18 $A \cap B$ is a subgroup of G. Clearly A is a subgroup of AB and

$A \cap B$ is a subgroup of B. Since $A \subset AB \subset G$, A is normal in AB by Theorem 6-4, and AB/A is defined.

To show that $A \cap B$ is normal in B, let $x \in A \cap B$ and $b \in B$. By the normality of A there exists $y \in A$ such that $xb = by$. Then $b^{-1}xb = b^{-1}by = y$; thus $b^{-1}xb \in A$. Since b, b^{-1}, and x are all in B, $b^{-1}xb \in B$. Therefore $b^{-1}(A \cap B)b \subset A \cap B$, and $A \cap B$ is shown to be normal in B by Theorem 3-7. Thus $B/A \cap B$ is defined.

Let us define a mapping $\alpha : B \to AB/A$ by $b\alpha = bA$ for all $b \in B$. Now an element of AB/A has the form abA with $a \in A$ and $b \in B$. Since A is normal in G, there exists $a_1 \in A$ such that $ab = ba_1$. Thus $abA = b(a_1A) = bA$, and α is onto. If $b_1,b_2 \in B$, then

$$(b_1b_2)\alpha = b_1b_2A = (b_1A)(b_2A) = (b_1\alpha)(b_2\alpha)$$

and α is a homomorphism. The kernel of α is $K = \{b \in B \mid b\alpha = A\}$ and $AB/A \cong B/K$ by Theorem 3-9. But $b\alpha = bA$, so $b \in K$ iff $bA = A$, which implies that $b \in A$. Therefore, $b \in K$ iff $b \in B$ and $b \in A$. Thus $K = A \cap B$ and the proof is complete. ||

This result will be helpful in the next section in the particular case $A \cap B = \{e\}$ and $AB = G$. In this case the conclusion reads

$$G/A \cong B$$

With all these results about homomorphisms and normal subgroups it should be clear that if we know the group G and the normal subgroup H, then G/H is completely determined. However, it is possible to know the normal subgroup H and the factor group G/H and still not know the structure of G. For example, if $G_1 = S_3$ and

$$G_2 = \{e,a,a^2,a^3,a^4,a^5\}$$

is the cyclic group of order 6, if $H_1 = \{(1),(123),(132)\} \subset G_1$ and

$$H_2 = \{e,a^2,a^4\} \subset G_2$$

then H_1 is normal in G_1, H_2 is normal in G_2, H_1 and H_2 are cyclic groups of order 3, and G_1/H_1 and G_2/H_2 are cyclic groups of order 2. Thus $H_1 \cong H_2$ and $G_1/H_1 \cong G_2/H_2$, but G_1 is not isomorphic to G_2.

Exercises 6-6 to 6-11

6-6. Prove Exercises 3-48 and 3-49 (if not done earlier).

6-7. Prove Theorem 6-4.

6-8. In the cyclic group $[a]$ of order 30 consider subgroups $H = [a^3]$ and $K = [a^{15}]$. Identify and characterize the corresponding structures mentioned in Corollary 6-5*a*.

6-9. In Table 3 of Appendix I (symmetries of the square) $\{I,R^2\}$ is a normal subgroup of the group S. Define a homomorphism α of S onto a group $\bar{G}$ of order 4 with kernel $\{I,R^2\}$. Find a normal subgroup $\bar{H}$ of order 2 in $\bar{G}$ and verify Theorems 6-3 and 6-5 for this case.

6-10. Let G be the complex numbers C with operation addition. Let $\bar{G}$ be the real numbers with operation addition. Let

$$H = \{a + bi \mid a,b \in J\}$$

Let $\alpha : G \to \bar{G}$ be defined by $(a + bi)\alpha = b$. Does Theorem 6-5 apply? (Justify your answer.) What is $\bar{H} = H\alpha$?

6-11. Let G be the group $\{a + bi \mid a,b \in J\}$ with operation addition. Let $A = \{a + bi \mid a \text{ is a multiple of 4 and } b = 0\}$ and

$$B = \{a + bi \mid a \text{ and } b \text{ are multiples of } 6\}$$

Verify Theorem 6-6 for this case. (Note that AB becomes $A + B$ here.)

6-B DIRECT PRODUCTS

The idea of this section was first discussed in Section 2-C, where we constructed a group from the cartesian product $G \times H$ of two groups G and H.

Definition 6-2

The **external direct product** of groups $\langle G,\circ\rangle$ and $\langle H,*\rangle$ is the system $\langle G \times H, \cdot\rangle$ such that, for all $(g_1,h_1),(g_2,h_2) \in G \times H$,

$$(g_1,h_1) \cdot (g_2,h_2) = (g_1 \circ g_2, h_1 * h_2)$$

This group will be denoted by $G \times H$.

We shall dispense with the operation symbols $\circ$, $*$, and $\cdot$ in what follows and use juxtaposition for all the group operations.

Theorem 6-7

For any two groups G and H, $G \times H$ is a group. (This is Exercise 2-49.)

PROOF. Definition 6-2 yields elements of $G \times H$, so the closure property holds. Associativity follows from the associativity of the operations in G and H. Also the identity of $G \times H$ is (e,e') where e is the identity of G and e' the identity of H, and $(g,h)^{-1} = (g^{-1},h^{-1})$. ||

Theorem 6-8

In $G \times H$, there are subgroups G_1 and H_1 such that $G_1 \cong G$ and $H_1 \cong H$.

PROOF. Let $G_1 = G \times \{e'\}$ where e' is the identity of H. Then the mapping $\alpha : g \to (g,e')$ is an isomorphism of G onto G_1. Clearly this mapping is one-to-one and onto and

$$(g_1g_2)\alpha = (g_1g_2,e') = (g_1,e')(g_2,e') = (g_1\alpha)(g_2\alpha)$$

The proof that H is isomorphic to $\{e\} \times H = H_1$, with e the identity of G, is similar. ||

Theorem 6-9

The subgroups G_1 and H_1 of Theorem 6-8 are normal subgroups of $G \times H$ and

$$G \times H/H_1 \cong G \qquad \text{and} \qquad G \times H/G_1 \cong H$$

PROOF. To show that H_1 is normal in $G \times H$, let $(g,h) \in G \times H$ and $(e,h') \in H_1$. Then

$$(g,h)^{-1}(e,h')(g,h) = (g^{-1}eg,h^{-1}h'h) = (e,h^{-1}h'h)$$

so that $(g,h)^{-1}H_1(g,h) \subset H_1$ and H_1 is normal by Theorem 3-7. Clearly $G_1H_1 = \{(g,e')(e,h) \mid g \in G \text{ and } h \in H\}$ is equal to $G \times H$. Also $G_1 \cap H_1 = \{(e,e')\}$. By Theorem 6-6 $G \times H/H_1 \cong G_1/\{(e,e')\}$. But $G_1/\{(e,e')\} \cong G_1 \cong G$. Thus

$$G \times H/H_1 \cong G$$

The proof for G_1 and $G \times H/G_1$ is similar. ||

All of these ideas can be extended to external direct products of three or more (even infinitely many) groups. The exercises will request a development of the theory in these cases.

One of the important applications of this idea is to the structure of vector spaces (as we shall see in Chapter 7). If G and H are both the group $\langle R,+\rangle$, then the *external direct sum* ("sum" because of the addition operation) $R \oplus R$ is the group whose elements are ordered pairs of real numbers and whose operation (also denoted by $+$) is defined by

$$(x_1,y_1) + (x_2,y_2) = (x_1 + x_2,\ y_1 + y_2)$$

What the theorems above are saying in this case is that $R \oplus R$ is a group (Theorem 6-7), that the x axis $\{(x,0) \mid x \in R\}$ and y axis $\{(0,y) \mid y \in R\}$ are subgroups of it (Theorem 6-8), and that the x axis (real numbers) is isomorphic to the system obtained as cosets of the y axis (the cosets are vertical lines).

Exercises 6-12 to 6-17

6-12. Show that every element of G_1 commutes with every element of H_1 in $G \times H$ and that $G \times H \cong H \times G$.

6-13. Define the external direct product $G \times H \times K$ of three groups G, H, and K. Prove that $G \times H \times K$ is isomorphic to $(G \times H) \times K$ (which is defined by Definition 6-2 used twice).

6-14. Show that $G \times H$ is isomorphic to a normal subgroup M of $G \times H \times K$ and that $G \times H \times K/M \cong K$.

6-15. Let C be the complex numbers. Show that $\langle C,+\rangle$ is isomorphic to $R \oplus R$.

6-16. Prove the unproved half of Theorem 6-9.

6-17. Let F be the set of all mappings with domain $\{0,1,2,3,4\}$ and range a subset of J. Show that $\langle F,+\rangle$ can be identified with $J \oplus J \oplus J \oplus J \oplus J$.

▲

We have observed, or can observe quickly, four properties of the external direct product $G \times H$ of two groups:

1. G and H are isomorphic to normal subgroups G_1 and H_1 of $G \times H$.
2. $G_1 \cap H_1 = \{(e,e')\}$, the identity subgroup of $G \times H$.
3. If $x \in G_1$ and $y \in H_1$, then $xy = yx$ in $G \times H$.
4. Every $(g,h) \in G \times H$ can be expressed uniquely as a product of an element of G_1 and an element of H_1 [as $(g,h) = (g,e')(e,h)$].

To capitalize on these ideas, we take an entirely different point of view. We shall assume that a group E with identity e is given (corresponding to $G \times H$) and that E contains normal subgroups G and H such that $GH = E$ (in the sense of Definition 6-1) and $G \cap H = \{e\}$.

Definition 6-3

A group E is the **internal direct product** of its subgroups G and H iff

1. G and H are normal in E.
2. $E = GH$.
3. $G \cap H = \{e\}$.

This is not the only form of the definition. The theorem below gives an equivalent characterization of the internal direct product.

Theorem 6-10

A group E is the internal direct product of its subgroups G and H iff

4. For every $g \in G$ and every $h \in H$, $gh = hg$.
5. Every $x \in E$ is uniquely expressible as $x = gh$ with $g \in G$ and $h \in H$.

PROOF. To prove this theorem, we must show that the first three conditions imply the last two (4 and 5) and that the last two imply the first three.

If conditions 1, 2, and 3 hold, let $g \in G$ and $h \in H$. Since H is normal, $g^{-1}h^{-1}g \in H$; since G is normal, $h^{-1}gh \in G$. Thus

$$(g^{-1}h^{-1}g)h = g^{-1}(h^{-1}gh)$$

is in H and in G; but $G \cap H = \{e\}$. Therefore $g^{-1}h^{-1}gh = e$, and we have $gh = hg$, so that condition 4 holds. If $x \in E$ and $x = g_1h_1 = g_2h_2$, then $g_2^{-1}g_1 = h_2h_1^{-1}$, which is possible only if $g_2^{-1}g_1 = h_2h_1^{-1} = e$ because e is the only element common to G and H. This implies that $g_1 = g_2$ and $h_1 = h_2$, so that condition 5 holds.

Conversely, assume that conditions 4 and 5 hold. Clearly condition 5 implies condition 2. If there were an element $c \neq e$ in $G \cap H$, then $c = ce = ec$ would contradict the uniqueness provision of condition 5; thus condition 3 holds. Finally, to show that G (similarly, H) is normal, let $x \in E$ and $g \in G$. Then there exists $h \in H$ such that $x = gh$. Then

$$x^{-1}gx = (gh)^{-1}g(gh) = h^{-1}gh = h^{-1}hg = g$$

Thus $x^{-1}Gx \subset G$ and G is normal in E. ||

Theorem 6-11

If E is the internal direct product of its normal subgroups G and H, then $E \cong G \times H$.

This isomorphism, whose proof depends on identifying (g,h) with gh, permits us to translate between the languages of internal and external direct products. Some theorems may be easier to prove in one form than in the other.

Exercises 6-18 to 6-26

☆*6-18*. Prove Theorem 6-11.

6-19. If E is an internal direct product of G and H and if G' is a normal subgroup of G, then G' is normal in E.

6-20. Generalize Theorem 6-10 to an internal direct product

$$E = G_1G_2G_3$$

of three subgroups. [*Caution:* Condition 3 should state that $G_i \cap (G_jG_k) = \{e\}$ for all permutations $i,j,k = 1,2,3$.] Prove the resulting theorem.

6-21. If E is an internal direct product of G and H, G' a subgroup of G and H' a subgroup of H, then $G'H'$ is a subgroup of E and an internal direct product.

6-22. The *center* of $G \times H$ (Exercise 3-34) is the direct product of the centers of G and H.

6-23. Each coset of G in the internal direct product $E = GH$ contains exactly one element of H.

6-24. If E is an internal direct product of G and H and if A is a subgroup of E such that $G \subset A$, then $A = G(A \cap H)$.

6-25. If $m,n \in J_+$ and gcd $\{m,n\} = 1$, then $J_m \oplus J_n \cong J_{mn}$.

6-26. Show that $J_2 \oplus J_4$ and $J_3 \oplus J_3$ are not cyclic.

6-C CONJUGATES

In Theorem 3-6 we defined mappings of a group onto itself called inner automorphisms. These mappings were defined by

$$xI_a = a^{-1}xa \qquad \text{for all } x \in G$$

where a is a fixed element of G. In this section we consider various ideas—some new, some seen earlier—involving elements of the form $a^{-1}xa$. The first concept to be defined changes the orientation of inner automorphisms—where the element a was fixed.

Definition 6-4

Let G be a group and let $x,y \in G$. We say that x is **conjugate** to y and write

$$x \,\sigma\, y$$

iff *there exists* an element $g \in G$ such that $g^{-1}xg = y$.

Thus σ, the relation of conjugacy of elements of a group, can be rephrased to say that $x \,\sigma\, y$ iff x is mapped to y by *some* inner automorphism of G. This relation turns out to be an equivalence relation in G.

Theorem 6-12

Let G be a group with identity e. Then σ is an equivalence relation in G.

PROOF. The three properties of an equivalence relation follow quickly from the definition of σ. Since $e^{-1}xe = x$ for all $x \in G$, σ is reflexive. If $x \, \sigma \, y$, then there exists $g \in G$ such that $g^{-1}xg = y$; then $x = gyg^{-1} = (g^{-1})^{-1}yg^{-1}$ and $y \, \sigma \, x$; thus σ is symmetric.

Finally, if $x \, \sigma \, y$ and $y \, \sigma \, z$, then there exist $g,h \in G$ such that $g^{-1}xg = y$ and $h^{-1}yh = z$. Substituting the first into the second yields

$$h^{-1}(g^{-1}xg)h = (gh)^{-1}x(gh) = z$$

so σ is transitive in G. ||

Definition 6-5

The equivalence classes in G determined by the conjugacy relation σ are called the **conjugate classes** of G.

Of course, if G is an abelian group, conjugacy is equality and the conjugate classes are one-element sets. Thus our main concern in developing the theory of conjugates and related ideas is with non-abelian groups.

In Chapter 3, when considering normal subgroups H, we proved that $aH = Ha$ if, and only if, $a^{-1}Ha \subset H$ (Theorem 3-7). The theorem and corollary below contain a slightly altered version of this characterization of normality. The proofs are deferred to the exercises.

Theorem 6-13

Let G be a group. Then a subgroup H of G is normal in G iff every conjugate of every element of H is also an element of H.

Corollary 6-13*a*

A subgroup H of a group G is normal in G iff H is a union of conjugate classes.

The idea of products $a^{-1}xa$ can be extended to subsets of the group G. These considerations lead to a generalization of the kind of subgroups considered in Exercises 2-24 and 2-25.

Definition 6-6

Let G be a group, A a subset of G, and $g \in G$. Then the subset of G

$$g^{-1}Ag = \{g^{-1}ag \mid a \in A\}$$

is called a **conjugate** of A.

We establish first some simple properties of conjugate sets. The first thing to recall is that a *subgroup* H is normal in G iff $g^{-1}Hg \subset H$ for all $g \in G$ (and this condition implies that $g^{-1}Hg = H$).

Theorem 6-14

The conjugate of an intersection is the intersection of the conjugates. [More precisely, if $g \in G$, then $g^{-1}(\bigcap_{t \in I} A_t)g = \bigcap_{t \in I} (g^{-1}A_t g)$.]

PROOF. The element x is in $g^{-1}(\bigcap A_t)g$ iff $x = g^{-1}ag$ with $a \in A_t$ for every $t \in I$, iff $x \in g^{-1}A_t g$ for every $t \in I$, iff $x \in \bigcap(g^{-1}A_t g)$. ||

Corollary 6-14*a*

The intersection of a collection of normal subgroups of a group is normal.

Exercises 6-27 to 6-32

6-27. Let $A = \{(1),(12)\}$ and $B = \{(1),(123)\}$ in S_3. Find all conjugates of A and B in S_3.

6-28. Verify Corollary 6-13*a* for the subgroup $\{I,R,R^2,R^3\}$ of the group S of Table 3, Appendix I.

☆*6-29.* Prove Theorem 6-13 and Corollary 6-13*a*.

6-30. If H is a subgroup of G, then $g^{-1}Hg = H$ iff $g^{-1}Hg \subset H$.

6-31. Prove Corollary 6-14*a*.

☆*6-32.* A conjugate of a subgroup is a subgroup.

Definition 6-7

Let A be a nonempty subset of a group G. The **normalizer** of A is the set

$$N_A = \{x \mid xA = Ax\}$$

The **normalizer of an element** $a \in G$ is the set $N_a = N_{\{a\}}$.

Note that $N_A = G$ if A is a normal subgroup of G or if G is an abelian group. If $x \in N_A$, then, for each $a \in A$, there exists $b \in A$

such that $ax = xb$, and there exists $c \in A$ such that $xa = cx$. This implies that $N_a = \{x \mid xa = ax\}$ (the set of all elements of G that commute with a). Thus N_a can also be described as the set of all x mapped to themselves by the inner automorphism I_a—since $ax = xa$ iff $xI_a = a^{-1}xa = x$.

The normalizer N_a of an element a has already been shown to be a subgroup of the group G (Exercise 2-24). This is also true of the normalizer N_A of every subset A of G.

Theorem 6-15

If A is a nonempty subset of a group G, then N_A is a subgroup of G.

PROOF. This proof is designed to show that the definition of a group is satisfied by N_A. Of course, associativity is inherited from G.

Let $x,y \in N_A$ and let $v \in xyA$. Then $v = xya$ for some $a \in A$, and there exist $b,c \in A$ such that $ya = by$ and $xb = cx$. Thus $v \in xyA$ iff $v = xya = xby = cxy$ iff $v \in Axy$ (more simply but less rigorously, $xyA = xAy = Axy$). Thus $xy \in N_A$ if $x,y \in N_A$.

The identity e is in N_A since $eA = A = Ae$. If $x \in N_A$, then $xA = Ax$ and, for every $a \in A$, there exists $b \in A$ such that $xa = bx$. This implies that $ax^{-1} = x^{-1}b$ so that $Ax^{-1} \subset x^{-1}A$. The verification that $x^{-1}A \subset Ax^{-1}$ is similar. ||

One of the more interesting results concerning conjugates relates the number of conjugates of an element x in G to the index (number of left—or right—cosets) of the subgroup N_x.

Theorem 6-16

Let x be an element of a group G. Then the number of conjugates of x is the index of N_x in G. (That is, there exists a one-to-one correspondence between the conjugates of x and the right cosets of N_x in G.)

PROOF. Let y be a conjugate of $x (y \sigma x)$ and let g and h be such that $g^{-1}xg = h^{-1}xh = y$. Then

$$(gh^{-1})^{-1}x(gh^{-1}) = h(g^{-1}xg)h^{-1} = hyh^{-1} = x$$

Thus $x(gh^{-1}) = (gh^{-1})x$ and $gh^{-1} \in N_x$. By the analog of Theorem 3-3 for right cosets (see Exercise 3-25), $N_xg = N_xh$ iff $gh^{-1} \in N_x$.

Thus all elements z which map x to y by $z^{-1}xz = y$ are in the same right coset of N_x in G.

Conversely, if $h \in N_x g$, then $h = ng$ with $nx = xn$. If $g^{-1}xg = y$, then

$$h^{-1}xh = g^{-1}(n^{-1}xn)g = g^{-1}xg = y$$

Thus all members of the same right coset of N_x map x to the same element y.

Thus there is a one-to-one correspondence between the right cosets of N_x in G and the conjugates y of x, and the theorem is proved. ||

Corollary 6-16*a*

The number of elements of a conjugate class of a finite group divides the order of the group.

PROOF. Theorem 6-16 and Theorem 3-5 (Lagrange's theorem). ||

Theorem 6-17

If A is a conjugate class of elements of a group G and if n is a positive integer, then

$$A^n = \{a^n \mid a \in A\}$$

is also a conjugate class.

PROOF. Let $n > 0$ and let $a \sigma b$. Then there exists $g \in G$ such that $g^{-1}ag = b$. Then b^n is a product of n factors $g^{-1}ag$ and

$$b^n = (g^{-1}ag)(g^{-1}ag) \cdots (g^{-1}ag) = g^{-1}a^n g$$

Thus if $a \sigma b$, then $a^n \sigma b^n$.

Conversely, if $c \sigma a^n$, then, for some $g \in G$,

$$c = g^{-1}a^n g = g^{-1}(agg^{-1})^n g = (g^{-1}ag)^n$$

so c is the nth power of a conjugate of a. ||

Exercises 6-33 to 6-39

6-33. If $A \subset B$ in a group G, what is the relationship between N_A and N_B? Prove your answer.

6-34. Is $N_A = \{x \mid \text{for every } a \in A, \text{ there exists } b \in A \text{ such that } ax = xb\}$?

6-35. Show that $N_A = A$ if $A = \{(1),(12)\}$ in S_3.

6-36. Show that $N_a \subset N_{[a]}$ (where $[a]$ is the cyclic subgroup generated by a). Show that $N_{[a]} \subset N_a$ may not hold.

☆*6-37.* The product AB of two conjugate classes A and B of G is a union of conjugate classes.

6-38. Show that Theorem 6-17 holds also for $n \leq 0$.

6-39. Show by example that $N_A \neq \bigcap_{a \in A} N_a$. Does an inclusion relation always hold between these sets?

▲

We conclude this chapter with a couple of results concerning the *center* of a group as defined in exercises in Chapters 2 and 3.

Definition 6-8

The **center** C of a group G is the set

$$C = \{x \mid gx = xg \text{ for all } g \in G\}$$

of all elements which commute with every element of G.

In Exercise 2-25 it was shown that C is a subgroup of G, and in Exercise 3-34 C was shown to be normal in G. It is also true by the same argument needed there that every subgroup of C is also normal. Of course, C may not be a proper subgroup of G since $C = G$ iff G is an abelian group and $C = \{(1)\}$, the identity, if $G = S_3$. An example of a proper subgroup C is found in the group of symmetries of the square (Table 3 of Appendix I) where $C = \{I, R^2\}$.

Since C is a normal subgroup of G, it is possible to talk about the quotient group G/C. If G is abelian, then $G/C \cong \{e\}$ since $G = C$, but if G is not abelian, a fairly strong statement is possible.

Theorem 6-18

If G is a non-abelian group with center C, then G/C is not cyclic.

PROOF. Assume that G/C is cyclic with generator aC. Then $G/C = [aC]$ and every element $x \in G$ has the form $x = a^r c$ with $c \in C$.

If $x = a^r c$ and $y = a^s c_1$ are in G, then

$$xy = (a^r c)(a^s c_1) = (a^r a^s)(cc_1) = (a^s a^r)(c_1 c) = (a^s c_1)(a^r c) = yx$$

so G is abelian. ||

In spite of the fact that G/C is not cyclic if G is not abelian, G/C may be abelian. For example, in the symmetries of the square $C = \{I, R^2\}$ so that G/C has order 4. But all the groups of order 4 are abelian (Exercise 2-54).

Exercises 6-40 to 6-42

6-40. If x is the only element of order 2 in a group G, then x is in the center of G.

6-41. Let H and K be subgroups of a group G. Define $H \sigma K$ iff there exists $g \in G$ such that $H = g^{-1}Kg$. Show that σ is an equivalence relation on the set of all subgroups of G. Determine the equivalence classes if G is the group of symmetries of the square (Table 3 of Appendix I).

6-42. If A is a conjugate class of elements of G, then $\{N_a \mid a \in A\}$ is a class of conjugate subgroups.

7 VECTOR SPACES

It is the primary purpose of this chapter to provide an introduction to the concept of a vector space using as a starting point the information about groups and fields discussed in Chapters 2 to 4. This is not intended as a substitute for a course in linear algebra but as a device for showing the relationship of theorems proved earlier to the subject of vector spaces and for indicating a few of the variety of topics which can be considered in this context. Important ideas of dimension and basis and the algebra of matrices and determinants will not be considered, but convexity and linear functionals will receive brief treatment.

It is hoped the person who has studied linear algebra or who is soon to study linear algebra will obtain here some knowledge of the relationship of the study of algebraic systems to that area of mathematics.

7-A GROUPS WITH OPERATORS

In proceeding from the theory of groups to the theory of vector spaces it is convenient to establish an intermediate position. The choice of this position is dictated by the wish to consider eventually equations of the form

$$b(x + y) = bx + by$$

where x and y are elements of a group $\langle G,+\rangle$ and b is in some set F.

This formation of a "product" of an element of F with an element of G can be interpreted as a mapping $\alpha : F \times G \to G$ such that $(b,x)\alpha = bx$. It can also be interpreted as a *collection* of mappings $b_L : G \to G$ such that $xb_L = bx$ for each $b \in F$.

Definition 7-1

Let $\langle G,\circ\rangle$ be a group. Then a homomorphism of G to G is called an **endomorphism** of G.

Thus α is an endomorphism of G iff $\alpha : G \to G$ and

$$(x \circ y)\alpha = (x\alpha) \circ (y\alpha)$$

for all $x,y \in G$.

By comparing this definition with Definition 3-6, we see that an *automorphism* of a group G is a one-to-one and onto endomorphism of G. An example of a one-to-one endomorphism which is not onto and hence not an automorphism is given by the mapping $n \to 2n$ of the group $\langle J,+\rangle$. Since the homomorphic image of a group is a group (Theorem 3-1), the image of G under an endomorphism must be a subgroup of G.

Definition 7-2

A **group with operators** is a collection consisting of a group $\langle G,\circ\rangle$ and a set E of endomorphisms of G. (The elements of E are called *operators* on G.)

The set E can still be chosen rather arbitrarily. It may, or may not, include the identity mapping δ. If δ is the only operator on G considered, then the study of G with operator set $E = \{\delta\}$ is simply the study of G alone—ordinary group theory.

In addition ordinary ring theory can be considered as a special case of the theory of groups with operators—as we proceed to show. Consider a ring $\langle A,+,\ \rangle$. Let G be the additive group $\langle A,+\rangle$ and, for each $a \in A$, define the mappings a_L and a_R by

$$xa_L = ax \qquad \text{and} \qquad xa_R = xa \qquad \text{for all } x \in A$$

In other words, a_L and a_R are, respectively, left and right multiplication by a. To say that a_L and a_R are endomorphisms of A is to say that the distributive laws hold. For example

$$(x + y)a_L = a(x + y) = ax + ay = xa_L + ya_L$$

Thus a ring may be described as a commutative group with operator set

$$E = \{a_L \mid a \in A\} \cup \{a_R \mid a \in A\}$$

The associative property of multiplication is equivalent to the associativity of composition of mappings (Theorem 1-2).

Exercises 7-1 to 7-10

7-1. Let A be a ring. Let $a,b \in A$. Show of the composites that $a_Rb_R = (ab)_R$. What is the corresponding statement for a_Lb_L? Verify associativity of multiplication using associativity of composition.

7-2 to 7-6. (**Admissible subgroups**) An *admissible subgroup* of a group G with operator set E is a subgroup H of G which is mapped into itself by every element of E.

7-2. Show that G and $\{e\}$ are admissible subgroups of G.

7-3. Let E be the set of all inner automorphisms of G (Theorem 3-6). What subgroups of G are admissible?

7-4. Let $\langle A,+\rangle$ be the additive group of a ring. Let the operator set E consist of all left and right multiplications a_L and a_R defined by elements of A. What subgroups of G are admissible?

7-5. The intersection of any family of admissible subgroups of a group with operators is an admissible subgroup.

7-6. State and prove a fundamental theorem of homomorphisms of groups with operators similar to Theorem 3-9. (You will need to consider admissible normal subgroups instead of all normal subgroups and limit yourself to homomorphisms which commute with every element of the operator set.)

7-7 to 7-10. (***For calculus scholars***) Let $C[0,1]$ denote the set of all continuous functions with domain the closed interval $[0,1]$ with operation addition. Let $a \in [0,1]$. Consider the operator S_a defined for each $f \in C[0,1]$ by $fS_a = \int_a^x f = \int_a^x f(t)\,dt$ (using conventional calculus notation for integrands).

7-7. Show that $E = \{S_a \mid a \in [0,1]\}$ is a set of operators on $C[0,1]$.

7-8. Let $f(x) = x + 1$ on $[0,1]$. Find the images fS_0, $fS_{1/2}$, $fS_0S_{1/2}$, and $fS_{1/2}S_0$.

7-9. Show that no S_a is a mapping of $C[0,1]$ onto $C[0,1]$.

7-10. Show that the derivative operator is not an endomorphism of $C[0,1]$.

7-B VECTOR SPACES

A vector space is a very special kind of group with operators. In particular, the operator set has some significant algebraic structure—it is a field—and the field operations and the group operation are related by certain equations. As indicated in the last section, a vector space can be defined in a rather complicated notation using concepts of mappings from $F \times G$ to G. We shall adopt a simpler form of the concept and speak of a "multiplication" of elements of F by elements of G.

Definition 7-3

An abelian group $\langle G,+\rangle$ is called a **vector space** (or *linear space*) over a field $\langle F,+,\ \rangle$ iff there exists a mapping $\lambda : F \times G \to G$ (called *left* *scalar multiplication*), with the image $(a,x)\lambda$ denoted by ax for all $a \in F$ and all $x \in G$, such that for all $x,y \in G$ and all $a,b \in F$

V1: $(a + b)x = ax + bx$.
V2: $(ab)x = a(bx)$.
V3: $a(x + y) = ax + ay$.
V4: $1x = x$, where 1 is the unity of F.

The elements of F are called *scalars* and the elements of G are called *vectors* or *points*.

We are taking a slight risk in denoting both the operation in G and the addition in F by $+$, but it should be clear from the context which is intended. The question to answer is: "Are the terms separated by $+$ vectors or scalars?" For example, the $+$ on the left side of V1 is for F and the $+$ on the right side for G since $a,b \in F$ but $ax,bx \in G$. Condition V3 says that the elements of F act as a set of operators on G. Condition V4 is a "normalizing" condition, part of its purpose being to prevent ax from being the identity element of G for every a and x.

The letters a, b, c, d will normally be used to denote scalars, and the later letters of the alphabet will denote vectors.

Definition 7-4

Let $\langle G,+\rangle$ be a vector space over a field F. The identity element of G is called the **origin** of the vector space and will be denoted by θ.

The zero and unity elements of F will be denoted by 0 and 1, respectively.

The most important simple examples of vector spaces are those which are cartesian products of the scalar field. Let us consider first the real numbers. Since $\langle R,+\rangle$ is an abelian group, we can make a group of $R \times R$ following the outline of Sections 2-C and 6-B by defining

$$(x_1,y_1) + (x_2,y_2) = (x_1 + x_2,y_1 + y_2)$$

for all $x_1,x_2,y_1,y_2 \in R$. Also, $R \times R$ becomes a vector space over R if we define

$$a(x,y) = (ax,ay)$$

for all $a,x,y \in R$. The properties of a vector space are easily checked, and we get the vector space which is identified with the usual coordinate plane of analytic geometry.

This construction can be generalized to an arbitrary field F and the cartesian product of F used more than two times.

Theorem 7-1

Let $\langle F,+,\ \rangle$ be a field. Let

$$F^n = \{(a_1,a_2, \ldots ,a_n) \mid a_1,a_2, \ldots ,a_n \in F\}$$

Define addition and scalar multiplication in F^n by

$$(a_1,a_2, \ldots ,a_n) + (b_1,b_2, \ldots ,b_n) = (a_1 + b_1,a_2 + b_2, \ldots ,a_n + b_n)$$

and

$$a(a_1,a_2, \ldots ,a_n) = (aa_1,aa_2, \ldots ,aa_n)$$

with $a \in F$. Then F^n is a vector space over F.

The proof of this theorem is deferred to the exercises. Compare this with the construction following Theorem 6-9.

Thus R^2 and R^3 denote the usual two- and three-dimensional coordinate spaces of analytic geometry. (Actually, the concept of a *distance* between vectors—or points—in R^2 and R^3 is also necessary to produce the familiar coordinate spaces in all their utility.) Theorem 7-1 says that $J_3{}^2$ is a vector space over J_3, and we have pictured it in Figure 7-1. Notice that the subset $\{(0,0),(2,1),(1,2)\}$ of $J_3{}^2$ is closed under both addition and scalar multiplication. Such subsets are called subspaces.

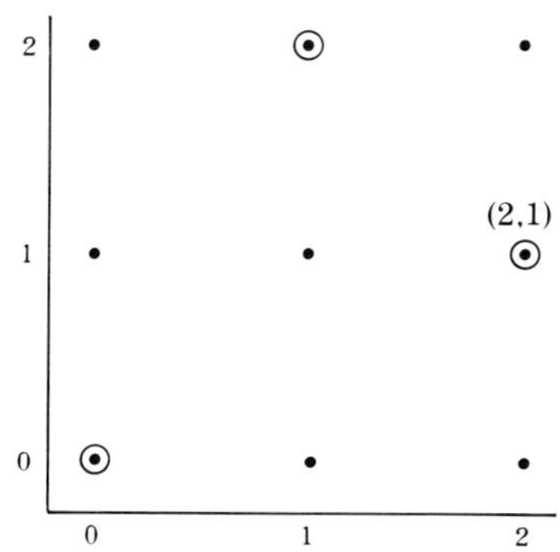

Figure 7-1 ***$J_3{}^2$ and a subspace.***

Definition 7-5

Let G be a vector space over a field F. Then a nonempty subset H of G is called a **subspace** of G iff H is closed with respect to addition and scalar multiplication. (Thus H is a subspace of G iff $ax + by \in H$ for all $x,y \in H$ and all $a,b \in F$.)

Exercises 7-11 to 7-22

7-11 to 7-16. **(Cancellation laws and related matters)**

☆*7-11.* For all $x \in G$ and all $a \in F$, $0x = a\theta = \theta$.

7-12. If $ax = \theta$, then $a = 0$ or $x = \theta$.

7-13. If $a,b \in F$ and $x \neq \theta$ in G and if $ax = bx$, then $a = b$.

7-14. If $a \neq 0$ in F and $x,y \in G$ and if $ax = ay$, then $x = y$.

7-15. Show that condition V1 defines an isomorphism of $\langle F,+\rangle$ to $\langle G,+\rangle$ if $x \neq \theta$ and x is fixed in G.

7-16. Let $a \neq 0$ in F. Show that the mapping $x \rightarrow ax$ is an automorphism of the group G.

7-17. Prove Theorem 7-1.

7-18. a. Let x be a fixed element of a vector space G over the field F. Show that $\{ax \mid a \in F\}$ is a subspace of G.

☆b. Let H be a subspace of G and $y \notin H$. Show that

$$K = \{ay + x \mid x \in H \text{ and } a \in F\}$$

is a subspace of G.

c. Describe geometrically in R^3 the subspaces obtained by the constructions of parts a and b above.

7-19. Let $x = (4,2)$ be an element of the vector space $J_7{}^2$. Exhibit the subspace $\{ax \mid a \in J_7\}$ on a coordinate grid like that of Figure 7-1.

7-20. Call a subspace of the form $\{ax \mid a \in F\}$, $x \neq \theta$, an *origin line* in F^n. (This will be a line through the origin in the usual sense in R^n over R.) Describe origin lines in $J_p{}^2$ over J_p, where p is a prime.

7-21. Find the smallest subspace of $J_3{}^4$ containing $(1,1,1,0)$ and $(2,0,0,1)$.

7-22. Show that the set $F[x]$ of all polynomials over a field F is a vector space over F. Compare Theorem 7-1 and Section 5-B.

▲

Let us now consider the relationship of the theory of groups developed in Chapters 2 and 3 (and 6) to the study of vector spaces and the geometry of R^2 and R^3. Roughly speaking, a vector space is an abelian group $\langle G,+\rangle$ with a scalar multiplication, and we hope to see the relationship of earlier theorems to $\langle G,+\rangle$.

The cancellation laws, part 1 of Theorem 2-1 and part 2 of Theorem 2-2, serve to complete the list of such properties in a vector space. These and Exercises 7-13 and 7-14 show that all forms of cancellation not involving θ and 0 are possible.

The solvability of group equations (Theorem 2-4) which says that the equation $u + x = v$ is satisfied by a unique $x \in G$ can be interpreted as saying that it is possible to get from any point u in a vector space G to any other point v with a single addition—in one step, that is. Thus any point is directly accessible from any other point.

If G and L are vector spaces over the same field F, then a homomorphism α from G to L would, by definition, satisfy the equation

$$(x + y)\alpha = x\alpha + y\alpha$$

for all $x,y \in G$. Theorem 3-1 tells us that the range of α is a *subgroup* $\bar{G}$ of L. There is still the question of whether or not $\bar{G}$ is a *subspace* of L. This is basically a question of whether or not α preserves scalar multiplication.

Definition 7-6

Let G and L be vector spaces over F. A mapping $\alpha : G \to L$ is called a **linear mapping** iff $(x + y)\alpha = x\alpha + y\alpha$ for all $x,y \in G$ and $(ax)\alpha = a(x\alpha)$ for all $x \in G$ and all $a \in F$.

Theorem 7-2

If α is a linear mapping from G to L, where G and L are vector spaces over F, then $\bar{G} = \text{rn}\ (\alpha)$ is a subspace of L.

PROOF. Let $u,v \in \bar{G}$ and $a,b \in F$. We hope to show that

$$au + bv \in \bar{G}$$

By hypothesis, there exist $x,y \in G$ such that $x\alpha = u$ and $y\alpha = v$. Then

$$au + bv = a(x\alpha) + b(y\alpha) = (ax)\alpha + (by)\alpha = (ax + by)\alpha$$

Since $ax + by \in G$, we have $au + bv \in \bar{G}$. ||

Every linear mapping between vector spaces is, by definition, a group homomorphism. It will be an isomorphism whenever Exercise 3-47 is satisfied—that is, whenever only the origin of G maps to the origin of L. A linear isomorphism thus effectively preserves all of the vector structure of G.

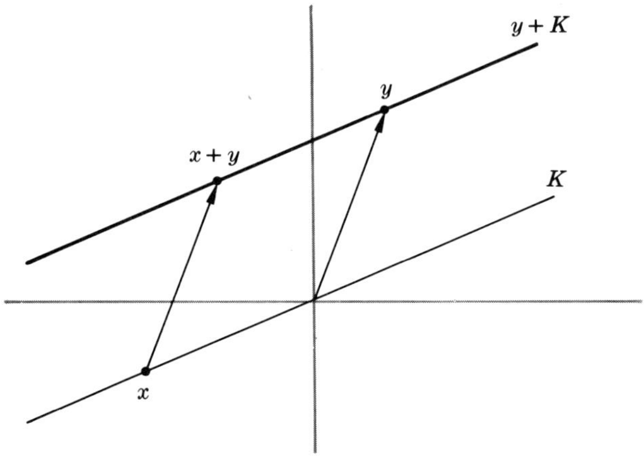

***Figure 7-2** A coset of a subspace K of R^2.*

If $\alpha : G \to L$ is a linear mapping, then not only is its kernel

$$K = \{x \in G \mid x\alpha = \theta'\}$$

where θ' is the origin in L, a normal subgroup of G (Theorem 3-9a), but since

$$(ax)\alpha = a(x\alpha) = a\theta' = \theta'$$

for all $a \in F$ and all $x \in K$, K is a subspace of G. Thus the cosets of K have the form $y + K$ for $y \in G$. Also we may define

$$a(y + K) = ay + K$$

which is justified by the computation

$$a(y + k) = ay + ak \in ay + K$$

Thus the factor group G/K (Definition 3-8) becomes a vector space. In a manner similar to the proof of Theorem 3-9*b* it can be shown that G/K is linearly isomorphic to $\bar{G} = \text{rn}\ (\alpha)$. The construction of G/K above can be accomplished for any subspace K of G without the introduction of a linear mapping α.

The geometry of cosets in R^2 and R^3 is fairly simple. If K is a line through the origin in R^2, then $y + K$ is a *translate* of that line—a line parallel to K and passing through y (see Figure 7-2). Similar statements hold if K is a line or plane through the origin in R^3.

Exercises 7-23 to 7-27

7-23. The intersection of two subspaces of a vector space G over F is a subspace of G.

7-24. If H and K are subspaces of G over F, then

$$H + K = \{h + k \mid h \in H \text{ and } k \in K\}$$

is a subspace of G.

7-25. Discuss the geometry of the analogs of Theorems 6-5 and 6-6 in R^3.

7-26. Describe geometrically the cosets of origin lines (in the sense of Exercise 7-20) in $J_p{}^2$.

☆*7-27.* Let G and L be vector spaces over F. Let $\mathfrak{L}(G,L)$ denote the set of all linear mappings from G to L. Show that $\mathfrak{L}(G,L)$ can be made into a vector space over F.

7-C CONVEXITY

This section is included to give the interested reader an opportunity to develop facility in working with vector spaces at a point of contact with ideas of considerable importance in modern analysis and geometry. Convex sets play an important part in several advanced areas of mathematics in which active research is being conducted. There is also a well-developed geometric theory of convex sets in R^n, especially in R^2 and R^3.

Definition 7-7

Let A and B be subsets of a vector space V over the real numbers R, and let $x \in V$ and $a \in R$. Then

1. $A + B = \{u + v \mid u \in A \text{ and } v \in B\}$.
2. $aA = \{au \mid u \in A\}$.
3. $A - B = A + (-1)B$.
4. $x + A = \{x\} + A$.

The set $x + A$ is called the **translate** of A by x. A translate of a subspace of V is called a **flat.**

This definition includes the basic notation we will need to discuss convexity. Note that the arithmetic of sets does not conform to all the usual rules. For example, $A + A \neq 2A$ in a vector space over R. To show this, let $A = \{x,y\}$ be a set of two vectors. Then $2A = \{2x,2y\}$, but $A + A = \{2x,2y,x + y\}$. However, it can be shown that $2A \subset A + A$ in all cases.

Definition 7-8

A **line** in V is a set of the form

$$\{ax + (1 - a)y \mid a \in R\}$$

where x and y are distinct fixed elements of V.

This definition yields the usual geometric idea of a line in R^2 and R^3, and is actually obtained by analogy with the point-of-division and midpoint formulas of analytic geometry.

Theorem 7-3

A line in V is a flat; in particular, a line is a translate of a subspace of the form $\{az \mid a \in R\}$ where $z \neq \theta$ in V.

PROOF. Let x and y be distinct elements of V and let

$$L = \{ax + (1 - a)y \mid a \in R\}$$

be a line in V. Let $H = L - y$. Since $y \in L$ (for $a = 0$), $\theta = y - y$ is in H. But $ax + (1 - a)y$ is in L iff $ax + (1 - a)y - y = a(x - y)$ is in H. Thus $H = \{a(x - y) \mid a \in R\}$, which is a subspace of V by

Exercise 7-18*a*. Since $L = y + H$ and $x - y \neq \theta$, the proof is complete. ||

The work of this section up to this point actually holds in a vector space over any field, although it has been formulated for vector spaces over the field R of real numbers. Since the following ideas involve inequalities, we shall continue to use R, although now any *ordered* field will be appropriate. We wish to consider the interval, or line segment, determined by the elements x and y—that is, the elements on the line determined by x and y which are "between" x and y.

Definition 7-9

Let V be a vector space over R. Let x and y be distinct elements of V. The **line segment** determined by x and y is the set

$$\overline{xy} = \{ax + (1 - a)y \mid 0 \leq a \leq 1\}$$

A subset K of V is **convex** iff, for all $x,y \in K$, $\overline{xy} \subset K$. [Thus K is convex iff $aK + (1 - a)K \subset K$ whenever $0 \leq a \leq 1$.]

Convex sets are well behaved with respect to set and vector operations. The basic properties are listed in the following theorems. Note that $\varnothing$ and $\{x\}$ are convex sets.

Theorem 7-4

If A and B are convex sets in a vector space V over R and if $b \in R$, then $A + B$ and bA are convex.

PROOF. If $u,v \in A + B$, then $u = x_1 + y_1$ and $v = x_2 + y_2$ with $x_1,x_2 \in A$ and $y_1,y_2 \in B$. If we consider $au + (1 - a)v$ with $0 \leq a \leq 1$, we find

$$\begin{aligned} au + (1 - a)v &= a(x_1 + y_1) + (1 - a)(x_2 + y_2) \\ &= [ax_1 + (1 - a)x_2] + [ay_1 + (1 - a)y_2] \end{aligned}$$

Since A is convex, $ax_1 + (1 - a)x_2$ is in A. Since B is convex, $ay_1 + (1 - a)y_2$ is in B. Thus $au + (1 - a)v \in A + B$, and $A + B$ is convex. The proof of the second part is deferred to the exercises. ||

Corollary 7-4*a*

If A is convex, then every translate $y + A$ of A is convex.

PROOF. The set $y + A = \{y\} + A$ is the sum of convex sets. ||

Corollary 7-4b

If a set A is convex, then $A + A = 2A$.

Theorem 7-5

The intersection of any collection of convex sets is convex.

Exercises 7-28 to 7-34

7-28. Let A be the union of the intervals $[0,1]$ on the x and y axes in R^2. Let B be the square with vertices $(0,0)$, $(0,1)$, $(1,0)$, and $(1,1)$ (just the square, not its interior). Let C be the circle $\{(x,y) \mid x^2 + y^2 = 1\}$. Find and graph in R^2:

a. $A + A$ b. $2A$ c. $A + B$

d. $B + B$ e. $2C$ f. $B + C$

7-29. Complete the proof of Theorem 7-4.

7-30. a. Prove that $2A \subset A + A$ for all A in a vector space V over R.

b. Prove Corollary 7-4b.

c. Let $a,b \in R$ such that $a > 0$ and $b > 0$. Prove that $aA + bA = (a + b)A$ if A is convex.

7-31. Prove Theorem 7-5.

7-32. When is $B = A + B - A$? Is $B \subset A + B - A$?

7-33. Every flat contains the line determined by each pair of its elements.

7-34. Can the sum of two nonconvex sets be convex? Can the sum of a convex set and a nonconvex set be convex?

▲

The next idea is based on Theorem 7-5. If we are given a set A in a vector space V over R, we can consider all the convex sets which contain A and form the intersection of those sets. This intersection will be a convex set contained in each of the given convex sets, and the intersection will contain A. It is, then, the smallest convex set containing A and may be interpreted as the set obtained by filling up the "indentations" in A.

Definition 7-10

Let A be a subset of a vector space V over R. The **convex hull** of A is the set

$$k(A) = \bigcap_{K \in S} K$$

where S is the collection of all convex sets containing A.

Two consequences of the definition are immediately apparent. First, $k[k(A)] = k(A)$—or, equivalently, B is convex iff $k(B) = B$. Second, if $A \subset B$, then $k(A) \subset k(B)$. Other properties of convex hulls can be obtained from the following characterization of $k(A)$.

Theorem 7-6

Let $\{x_1, x_2, \ldots, x_n\}$ be a finite subset of A. Let us call a sum of the form $\sum_{i=1}^{n} a_i x_i$ a *convex combination* of elements of A iff $a_i \geq 0$ for all i and $\sum_{i=1}^{n} a_i = 1$. Then the set $k(A)$ is the set of all convex combinations of elements of A.

DISCUSSION. The proof of this theorem involves some manipulations typical of convexity proofs. The difficulties encountered are largely traceable to the fact that convex combinations of elements of A do not all involve the same number of elements. The proof that the set of all convex combinations of elements of A is a subset of $k(A)$ will involve fixing the number of elements in a convex combination —and induction on that number.

PROOF. Let H be the set of all convex combinations of elements of A and let H_r be the set of all convex combinations of r elements of A. Then $H_1 = A$,

$$H_r = \left\{ \sum_{i=1}^{r} a_i x_i \mid x_i \in A, \text{ and } a_i \geq 0 \text{ for all } i, \text{ and } \sum_{i=1}^{r} a_i = 1 \right\}$$

and

$$\bigcup_{r=1}^{\infty} H_r = H$$

Let K be any convex set containing A. We shall show $H \subset K$ [and thus $H \subset k(A)$] by showing $H_r \subset K$ for every r. Clearly $H_r \subset K$ for $r = 1$ and $r = 2$ since $A \subset K$ and K is convex.

Assume that $H_n \subset K$ and consider H_{n+1}. Let

$$x_1,x_2, \ldots ,x_n,x_{n+1} \in A$$

and consider $\sum_{i=1}^{n+1} a_i x_i$, a convex combination of $n + 1$ elements of A. Since $\sum_{i=1}^{n+1} a_i = 1$, let $\sum_{i=1}^{n} a_i = b = 1 - a_{n+1}$. Then

$$\frac{a_1}{b} x_1 + \frac{a_2}{b} x_2 + \cdots + \frac{a_n}{b} x_n$$

is in H since $\sum_{i=1}^{n} a_i/b = 1$, and

$$\sum_{i=1}^{n+1} a_i x_i = b \left[\sum_{i=1}^{n} \frac{a_i}{b} x_i \right] + a_{n+1} x_{n+1}$$

is in K since it is a convex combination of two elements of K. Since $H = \bigcup_{r=1}^{\infty} H_r \subset K$ and K is an arbitrary convex set containing A, H is a subset of every convex set containing A. Thus $H \subset k(A)$.

To show, conversely, that $k(A) \subset H$, we must show only that H is a convex set containing A. Clearly $A \subset H$ since $A = H_1$ and $H = \bigcup H_r$. If $x,y \in H$, then $x = \sum_{i=1}^{n} a_i x_i$ and $y = \sum_{i=1}^{m} b_i y_i$, where $a_i \geq 0$ and $b_i \geq 0$ for all i, $x_i,y_i \in A$ for all i, and $\sum_{i=1}^{n} a_i = \sum_{i=1}^{m} b_i = 1$. If $0 \leq a \leq 1$, then

$$\begin{aligned} ax + (1 - a)y &= a \sum_{i=1}^{n} a_i x_i + (1 - a) \sum_{i=1}^{m} b_i y_i \\ &= \sum_{i=1}^{n} (aa_i) x_i + \sum_{i=1}^{m} (1 - a) b_i y_i \end{aligned}$$

But this is a convex combination of $n + m$ elements of A since the sum of all the coefficients is 1. Thus H is convex, and the proof is complete. ||

Exercises 7-35 to 7-42

7-35. Show that $k(aA) = ak(A)$ for every $A \subset V$ and every $a \in R$.

☆*7-36.* Show that $k(A + B) = k(A) + k(B)$ for all $A,B \subset V$.

7-37 to 7-42. **(Cones)** A subset C of a vector space V over R is called a *cone* iff (1) for all $x,y \in C$ and all $a \geq 0$, $x + y$ and ax are in C, and (2) if $x, -x \in C$, then $x = \theta$. (Condition 2 says that C contains no line through θ.)

7-37. Every cone C is convex. Every convex set D such that $ax \in D$ for all $x \in D$ and all $a \geq 0$ and such that D contains no line through θ is a cone.

7-38. Let C and D be cones in V such that if $x \in C$ then $-x \notin D$ (unless $x = \theta$). Show that $C + D$ is a cone and that

$$C + D = k(C \cup D)$$

7-39. Let C be a cone in V. If $x,y \in V$, define $x \geq y$ iff $x - y \in C$.

a. Show that $\geq$ is reflexive and transitive in V.

b. Show that $\geq$ is *antisymmetric* in V. (That is, if $x \geq y$ and $y \geq x$, then $x = y$. A relation which is reflexive, antisymmetric, and transitive is called a *partial order*. The relation $\subset$ among subsets of a set X is another example of a partial order.)

c. If $x \geq y$ and $z \in V$, then $x + z \geq y + z$.

d. If $x \geq y$ in V and $a > 0$ in R, then $ax \geq ay$ in V. What happens if $a < 0$?

7-40. Let K be a convex set in V such that $\theta \notin K$. Show that $\{ax \mid x \in K \text{ and } a \geq 0\}$ is a cone in V.

7-41. A cone C is a *maximal cone* in V iff no cone properly contains C. Show that, if $C \cup (-C) = V$, then C is a maximal cone in V. (The converse is also true.)

7-42. If C is a cone in V, then $C - C$ is a subspace of V.

▲

The final topic of this chapter is the relation between linear mappings from V to the scalar field and subspaces of V. We have shown that if α is a linear mapping with domain V, then the kernel of α is a subspace of V. We are seeking a stronger result if the range of α is the scalar field R. (We shall continue to use R although the theorems which follow are valid for any scalar field F.)

Definition 7-11

Let V be a vector space over a field F. A mapping $\alpha : V \to F$ is called a **functional.**

Our concern will be with *linear functionals*—mappings from a vector space to its scalar field which satisfy the conditions

$$(x + y)\alpha = x\alpha + y\alpha$$

and $(ax)\alpha = a(x\alpha)$ of Definition 7-6.

In Exercise 7-18*b* it was asserted that if H is a subspace of V and $y \notin H$, then the set of all elements of the form $ay + x$ with $a \in F$

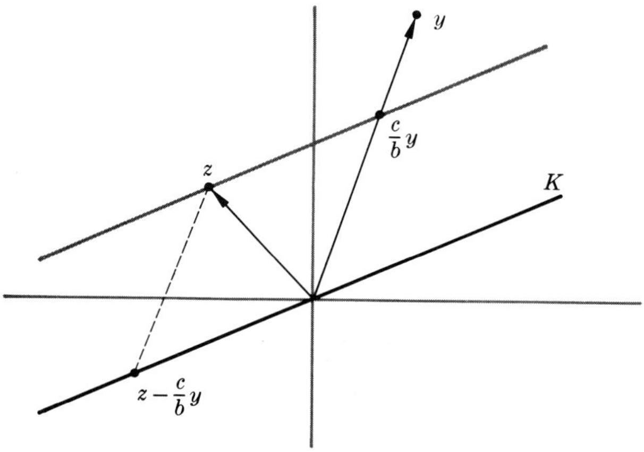

Figure 7-3 Geometry of the proof of Theorem 7-7.

and $x \in H$ is a subspace of V. Since any subspace containing y and H must contain all scalar multiples of y and all sums of pairs of elements, this subspace is the smallest subspace containing H and y. If the subspace $\{ay + x \mid a \in F \text{ and } x \in H\}$, $y \notin H$, happens to be V itself, then we shall call H a *maximal proper subspace* of V. A line through the origin in R^2 and a plane through the origin in R^3 are examples of maximal proper subspaces.

We shall say that a functional α is a *nonzero functional* iff there exists $x \in V$ such that $x\alpha \neq 0$.

Theorem 7-7

If α is a nonzero linear functional on a vector space V over R, then the kernel of α, $K = \{x \in V \mid x\alpha = 0\}$, is a maximal proper subspace of V.

PROOF. Since α is a nonzero linear functional on V, there is an element $y \in V$ such that $y\alpha = b \neq 0$. Then $y \notin K$. To show that K is a maximal proper subspace of V, we shall show that every $z \in V$ has the form $ay + x$ for some $a \in R$ and some $x \in K$. (The scheme of the proof is indicated by Figure 7-3.)

Let $z \in V$. Since dm $(\alpha) = V$, $z\alpha = c \in R$. Since

$$[(c/b)y]\alpha = (c/b)b = c$$

$[z - (c/b)y]\alpha = c - c = 0$ and $z - (c/b)y$ is an element of K. Thus

$$z = (c/b)y + [z - (c/b)y]$$

is the required representation of z in the form $ay + x$. ||

Theorem 7-8

(Converse of Theorem 7-7) If K is a maximal proper subspace of V, then there exists a nonzero linear functional α on V such that

$$K = \{x \mid x\alpha = 0\}$$

PROOF. Let $y \notin K$. Since K is maximal, every element of V can be written as $ay + x$ with $a \in R$ and $x \in K$. If $ay + x_1 = by + x_2$ with $x_1,x_2 \in K$, then $(a - b)y = x_2 - x_1$. Since K is a subspace, $x_2 - x_1 \in K$, which means that $y \in K$ unless $a = b$. Therefore, $a = b$ and $x_2 = x_1$, so the representation of an element of V as $ay + x$ is unique. Define a functional α by $(ay + x)\alpha = a$. The proofs that α is linear and has kernel K are left to the exercises. ||

In other words, we have established a correspondence between linear functionals on V and maximal proper subspaces of V. The correspondence is not one-to-one because every nonzero scalar multiple of a linear functional has the same kernel.

A translate $y + H$ of a maximal proper subspace H is called a *hyperplane*. Hyperplanes play an important part in the advanced theory of vector spaces over R. For example, many properties of convex subsets of vector spaces depend upon the existence of *supporting hyperplanes* (the analog of tangent planes to sets in R^3).

Corollary 7-8*a*

A subset L of a vector space V over R is a hyperplane iff there exist a nonzero linear functional α and a constant $c \in R$ such that

$$L = \{x \mid x\alpha = c\}$$

Other important theorems involve separating convex sets by a hyperplane. (Convex sets A and B are separated by a hyperplane $L = \{x \mid x\alpha = c\}$ iff $A \subset \{x \mid x\alpha \geq c\}$ and $B \subset \{x \mid x\alpha \leq c\}$—or vice versa.)

Exercises 7-43 to 7-47

7-43. Complete the proof of Theorem 7-8.

7-44. Prove Corollary 7-8*a*.

7-45 to 7-47 (***Minkowski functional—for analysis students***) Let V be a vector space over R. An element x of a subset A of V is a *core point* of A iff, for every $y \in V$, there exists $b_y > 0$ such that $ax \in A$ whenever $|a| < b_y$. (That is, x is a core point of A iff every line through x contains an open interval contained in A and containing x.) Let A be a convex set such that θ is a core point of A. The *Minkowski functional* of A is the functional β_A defined by

$$x\beta_A = \inf \{a \mid x \in aA\} = \inf \left\{a \mid \frac{1}{a}x \in A\right\}$$

for all $x \in V$.

7-45. Give an example of a set B in R^2 such that θ is a core point of B but θ is also a boundary point of B. (B will not be convex.)

7-46. Show that β_A has the following properties if A is a convex set with core point θ:

a. $(bx)\beta_A = b(x\beta_A)$ if $b \geq 0$.

☆b. $(x + y)\beta_A \leq x\beta_A + y\beta_A$ for all $x,y \in V$.

7-47. A functional γ on V over R is called a *norm* iff (1) $x\gamma = 0$ iff $x = \theta$, (2) $(ax)\gamma = |a|(x\gamma)$ for all a and x, and (3) $(x + y)\gamma \leq x\gamma + y\gamma$ for all x and y. (For example, absolute value is a norm on R^1 over R. Much of the importance of norms comes from the fact that $x\gamma$ can be used as a measure of distance of x from θ and $(x - y)\gamma$ is a measure of distance between x and y.) What additional conditions on A will make β_A a norm?

8 BOOLEAN ALGEBRA AND MATHEMATICAL PROOF

That branch of learning known as symbolic logic can be traced to the work of the English mathematician George Boole (1815–1864) who in 1854 published "An Investigation of the Laws of Thought on Which Are Founded the Mathematical Theories of Logic and Probabilities." The algebraic systems which evolved from Boole's work are known today as *Boolean algebras.* An introductory study of these is undertaken in the first section of this chapter. The second section develops very informally some concepts of logic useful in mathematical work and relates logic to Boolean algebra. Then, because much of this reminds one of the set theoretic ideas of Chapter 1, logic and Boolean algebra are related to set theory.

8-A BOOLEAN ALGEBRA

This section is devoted to establishing several properties of Boolean algebra. The system discussed here is unique in this book in that it is the only major one which does not depend on the group concept of Chapter 2.

Definition 8-1

A **Boolean algebra** is a system $\langle B, \vee, \wedge \rangle$ such that

B1: $\vee$ and $\wedge$ are commutative in B.
B2: There exist in B an identity s for $\wedge$ and an identity n for $\vee$.
B3: $\vee$ is distributive over $\wedge$ and $\wedge$ is distributive over $\vee$ in B.
B4: For each $x \in B$ there exists $x' \in B$ such that $x \vee x' = s$ and $x \wedge x' = n$.

The element x' is called a **complement** of x.

Names for the operations $\vee$ and $\wedge$ vary. One possible pair of names is "cup" and "cap"; another is "vee" and "wedge."

It is worth noting that each of the axioms B1 to B4 is actually a statement of two properties of the system and that the first property can be obtained from the second and the second from the first by substituting $\vee$ for $\wedge$, $\wedge$ for $\vee$, n for s, and s for n throughout. Because of this, whenever a theorem and proof are given for a Boolean algebra, one can obtain another theorem by making the above substitutions in the theorem and prove the new theorem by making the same substitutions in the proof. This property of a Boolean algebra is called the *principle of duality*. Thus to each theorem about our Boolean algebra B there corresponds a dual theorem (which occasionally might be the same theorem—see Exercise 8-6 for instance).

After proving two simple theorems about the Boolean algebra $\langle B, \vee, \wedge \rangle$, we shall summarize the basic properties of $\vee$ and $\wedge$ in table form. The first theorem and its dual are known as the *idempotent* laws.

Theorem 8-1

For each $a \in B$, $a \vee a = a$. (Dual theorem: $a \wedge a = a$.)

PROOF. The proof is a calculation, which proceeds as follows:

$$\begin{aligned} a &= a \vee n && \text{B2} \\ &= a \vee (a \wedge a') && \text{B4} \\ &= (a \vee a) \wedge (a \vee a') && \text{B3} \\ &= (a \vee a) \wedge s && \text{B4} \\ &= a \vee a && \text{B2} \quad \| \end{aligned}$$

Theorem 8-2

For each $a \in B$, $a \vee s = s$. (Dual theorem: $a \wedge n = n$.)

PROOF. Calculating again:

$$\begin{aligned} a \vee s &= (a \vee s) \wedge s && \text{B2} \\ &= s \wedge (a \vee s) && \text{B1} \\ &= (a \vee a') \wedge (a \vee s) && \text{B4} \\ &= a \vee (a' \wedge s) && \text{B3} \\ &= a \vee a' && \text{B2} \\ &= s && \text{B4} \quad \| \end{aligned}$$

In addition to the commutative and distributive properties, the axioms and theorems have already provided enough information to find the "$\vee$ product" and the "$\wedge$ product" of any pair of s, n, a, and a' where a is any element of B. The tables below summarize this information.

$\vee$	n	a	a'	s
n	n	a	a'	s
a	a	a	s	s
a'	a'	s	a'	s
s	s	s	s	s

$\wedge$	n	a	a'	s
n	n	n	n	n
a	n	a	n	a
a'	n	n	a'	a'
s	n	a	a'	s

It is a consequence of Exercise 1-57 and the commutativity of $\vee$ and $\wedge$ that the identity elements n and s are unique. It is also true that this uniqueness can be guaranteed also for the elements s and n in terms of the property stated in Theorem 8-2. We state these uniqueness properties in the following theorem and give proofs independent of Exercise 1-57. (We shall henceforth omit the reasons for steps in proofs that involve calculations and leave it to the reader to justify the steps in the proofs.)

Theorem 8-3

If x is an element of B such that $a \wedge x = x$ or $a \vee x = a$ for all $a \in B$, then $x = n$. (Dual theorem: If $a \vee y = y$ or $a \wedge y = a$ for all $a \in B$, then $y = s$.)

PROOF. If $a \wedge x = x$ for all $a \in B$, then, in particular, $n \wedge x = x$. But $n \wedge x = n$, so $x = n$.

If $a \vee x = a$ for all a, then $n \vee x = n$. But $n \vee x = x$, so $x = n$. ||

The next theorem and its dual are called the *absorption laws* for a Boolean algebra. They are of considerable use in the proofs which follow and are particularly valuable tools for calculation in a Boolean algebra.

Theorem 8-4

If $a,b \in B$, then $a \vee (a \wedge b) = a$. [Dual theorem: $a \wedge (a \vee b) = a$.]

PROOF. The following calculation suffices.

$$a \vee (a \wedge b) = (a \wedge s) \vee (a \wedge b) = a \wedge (s \vee b) = a \wedge s = a \quad ||$$

Corollary 8-4*a*

Let $a,b \in B$. Then $a \wedge b = a$ iff $a \vee b = b$.

The proof of this corollary is deferred to the exercises. The corollary is an example of a theorem which is its own dual.

Perhaps the reader has already observed that one of the systems which is a Boolean algebra is the system $\langle P(X), \cup, \cap \rangle$ of all subsets of a set X. If we identify $\cup$ with $\vee$, $\cap$ with $\wedge$, the empty set $\varnothing$ with n, X with s, and the set theoretic complement $\sim A = X \sim A$ with the Boolean complement, then the axioms B1 to B4 and the theorems are true statements about sets. We are already aware that it is not possible to "cancel" in set equations as easily as in equations in the number systems. For example, if $C = \{x\}$, $D = \{x,y\}$, and $E = \{y,z\}$, then $C \cup E = D \cup E$, but $C \neq D$. However, if we can get the appropriate combination of *two* equations to hold in a Boolean algebra, then the cancellation is permissible.

Theorem 8-5

Let $a,b,c \in B$. If $a \vee b = a \vee c$ and if $a \wedge b = a \wedge c$, then $b = c$. (Another self-dual theorem.)

PROOF. Calculating again:

$$
\begin{aligned}
b = b \wedge (a \vee b) = b \wedge (a \vee c) &= (b \wedge a) \vee (b \wedge c) \\
&= (a \wedge b) \vee (b \wedge c) = (a \wedge c) \vee (b \wedge c) \\
&= (a \vee b) \wedge c = (a \vee c) \wedge c = c \quad \|
\end{aligned}
$$

Corollary 8-5*a*

The complement a' of an element $a \in B$ is unique.

PROOF. Let a' and a_1 be complements of a in the sense of axiom B4. Then $a \vee a' = a \vee a_1 = s$ and $a \wedge a' = a \wedge a_1 = n$. By Theorem 8-5, $a' = a_1$. ||

Corollary 8-5*b*

For every $a \in B$, $(a')' = a$.

PROOF. Since $a' \wedge a = n$ and $a' \vee a = s$, a is a complement of a' in the sense of B4. Since complements are unique by Corollary 8-5*a*, $(a')' = a$. ||

The set operations $\cup$ and $\cap$ are associative. So far no mention has been made of the possibility of the associativity of $\vee$ and $\wedge$. These operations are associative, but the point is that this is a theorem which follows from the axioms and the other theorems. Actually it was some time after the formulation of axioms for a Boolean algebra that it was realized that an axiom of associativity was unnecessary and that associativity could be proved from the other axioms. Since we have not used associativity at any point previously, we can give the usual proof of the associative property without being accused of reasoning in circles.

Theorem 8-6

The operation $\vee$ is associative in B. (Dual theorem: The operation $\wedge$ is associative in B.)

PROOF. Let $a,b,c \in B$. Let $x = a \vee (b \vee c)$ and let

$$y = (a \vee b) \vee c$$

In order to show $x = y$, we first perform four calculations.

1. $a \wedge x = a \wedge [a \vee (b \vee c)] = a$ (Theorem 8-4).

2. $a \wedge y = a \wedge [(a \vee b) \vee c] = [a \wedge (a \vee b)] \vee (a \wedge c)$
$= a \vee (a \wedge c) = a$.

3. $a' \wedge x = a' \wedge [a \vee (b \vee c)] = (a' \wedge a) \vee [a' \wedge (b \vee c)]$
$= n \vee [a' \wedge (b \vee c)] = a' \wedge (b \vee c)$.

4. $a' \wedge y = a' \wedge [(a \vee b) \vee c] = [a' \wedge (a \vee b)] \vee (a' \wedge c)$
$= [(a' \wedge a) \vee (a' \wedge b)] \vee (a' \wedge c) = [n \vee (a' \wedge b)] \vee (a' \wedge c)$
$= (a' \wedge b) \vee (a' \wedge c) = a' \wedge (b \vee c)$.

Thus $a \wedge x = a \wedge y$ and $a' \wedge x = a' \wedge y$. From this we conclude that

$$\begin{aligned}(a \wedge x) \vee (a' \wedge x) &= (a \wedge y) \vee (a' \wedge y)\\ (a \vee a') \wedge x &= (a \vee a') \wedge y\\ s \wedge x &= s \wedge y\\ x &= y \quad \|\end{aligned}$$

As a final theorem we state De Morgan's laws in the context of a Boolean algebra. (Compare these with Exercises 1-7 and 1-14*a* and *b*.)

Theorem 8-7

Let $a,b \in B$. Then $(a \vee b)' = a' \wedge b'$. [Dual theorem: $(a \wedge b)' = a' \vee b'$.]

DISCUSSION. This theorem can be proved by showing that $a' \wedge b'$ is the complement of $a \vee b$—that is, $(a \vee b) \vee (a' \wedge b') = s$ and $(a \vee b) \wedge (a' \wedge b') = n$. The calculations are deferred to the exercises.

Exercises 8-1 to 8-7

8-1. Write out the proofs of the dual theorems to Theorems 8-1 and 8-2.

8-2. Prove Corollary 8-4*a*.

8-3. a. Prove Theorem 8-7, giving a reason to justify each step.

b. Let $a_1, a_2, \ldots, a_n \in B$. Show that

$$(a_1 \vee a_2 \vee \cdots \vee a_n)' = a_1' \wedge a_2' \wedge \cdots \wedge a_n'$$

(Use induction and the fact that all associations of $a_1 \vee a_2 \vee \cdots \vee a_n$ are equal because of Theorem 8-6. See Appendix II for this result.)

c. State the dual of part *b*.

8-4. Show that

$$a \vee (b_1 \wedge b_2 \wedge \cdots \wedge b_n) = (a \vee b_1) \wedge (a \vee b_2) \wedge \cdots \wedge (a \vee b_n)$$

in B. State the dual.

8-5. a. If $a,b \in B$, then

$$(a \wedge b) \vee (a' \wedge b) \vee (a \wedge b') \vee (a' \wedge b') = s$$

b. Discuss the analog of part *a* for set theory.

8-6. Prove $s' = n$ and $n' = s$ in B.

8-7. Prove

$$(a \wedge b) \vee (b \wedge c) \vee (c \wedge a) = (a \vee b) \wedge (b \vee c) \wedge (c \vee a)$$

in B.

8-B LOGIC AND PROOFS

Symbolic logic can be studied mathematically for its own sake. In fact, mathematical logic is a well-developed and active branch of mathematics. But for the mathematician not concerned with logic for its own sake, logic is a tool to be used in constructing proofs of theorems. It is the purpose of this section to develop a little logic from this second point of view and provide some intuitive answers to the question: "How do I go about the task of proving a theorem?"

The development of these answers begins with a study of the *form* of common mathematical statements. The study will be restricted to those statements which have the property that exactly one of the labels "true" and "false" applies to the statement. Such statements will be called *propositions,* and since we shall study the form of propositions rather than the propositions themselves, the principal objects of study will be *propositional forms* with *variables* p, q, r, etc.—things which become propositions when statements are substituted for the variables. Thus this study will focus on things like "If p, then q" rather than "If it rains, then the tennis courts will be closed." All propositional forms to be considered will be constructed from five basic forms, which will be introduced shortly.

The goal of this study is to discover some rules of reasoning which can be regarded as "safe" in the sense that conclusions reached using them do not depend on an element of luck—that is, the output is to be as reliable as the input. We begin by assigning *truth values* ("true" or "false") to the five basic propositional forms. We shall use "1" to symbolize "true" and "0" for "false."

Definition 8-2

The basic propositional forms are *p and q*, *p or q*, *if p then q* (also read as "p implies q" and symbolized by "$p \to q$"), *p if and only if q* (symbolized by "$p \leftrightarrow q$" or "*p iff q*"), and *not p*. These forms are defined in terms of their truth values corresponding to the truth values of their variables p and q by the following tables.

p	q	p *and* q	p *or* q	$p \to q$	$p \leftrightarrow q$
1	1	1	1	1	1
1	0	0	1	0	0
0	1	0	1	1	0
0	0	0	0	1	1

p	*not* p
1	0
0	1

In $p \to q$, p is called the **hypothesis** and q the **conclusion.**

To reduce the need for parentheses in writing complicated propositional forms, we shall declare that *not* refers only to the immediately following variable unless parentheses indicate otherwise.

Note that *or* is used in its inclusive sense of "one or the other or both" because if p is true and q is true, then p *or* q is true. The first two columns of the table are easily remembered: p *and* q is true iff p and q are both true, whereas p *or* q is true unless p and q are both false. The fourth column is also easy to remember because $p \leftrightarrow q$ is true iff p and q have the same truth value. It is the entries in the column for $p \to q$ that require some explanation.

Consider the proposition: "If he works, then he will succeed." If a person works and succeeds, we would be inclined to say that the proposition is true, but if he works and fails, then the proposition is false. However, to say that the proposition is true if the person does not work, we can best describe as merely the charitable view of not declaring a statement a lie unless it is proved to be so. However, the use of a false hypothesis in a true statement is fairly commonplace. Consider, for example, the proposition: "If anyone can solve that problem, then I'm a dirty bird." The person making such a statement clearly intends that the statement as a whole be found true without raising the question of whether he is, or is not, a dirty bird. His intent is to insinuate that the problem is unsolvable. In this case, and in general, $p \to q$ is true whenever p is false. For example, the empty set $\varnothing$ is a subset of every set since the statement "If $x \in \varnothing$, then $x \in A$" is true (because the hypothesis is false).

The process of determining the truth values of other propositional forms constructed from the five basic ones can best be accomplished by systematically constructing the form and keeping track of the truth values determined by Definition 8-2 for the component parts. We give two examples, p *and* ($p \to$ *not* q), and *not* p *or not* q.

p	q	*not* q	$p \to$ *not* q	p *and* ($p \to$ *not* q)
1	1	0	0	0
1	0	1	1	1
0	1	0	1	0
0	0	1	1	0

p	q	*not* p	*not* q	*not* p *or not* q
1	1	0	0	0
1	0	0	1	1
0	1	1	0	1
0	0	1	1	1

We are now in a position to define what we meant earlier by "safe" forms of reasoning.

Definition 8-3

A **tautology** is a propositional form which takes on the value 1 for every combination of values of its variables.

In using an argument which has the form of a tautology, a mathematician is guaranteeing that his argument will not contain flaws caused by variations in the truth of the propositions used as the things referred to by those propositions change. Thus mathematicians limit themselves to tautological forms in writing proofs.

It is easy to find some examples of tautologies involving $\leftrightarrow$. An examination of the last column in the first table above reveals that it is the opposite of the table for $p \rightarrow q$. Since $p \leftrightarrow q$ is true whenever p and q have the same truth value and since *not* reverses values, it follows that

$$not(p \rightarrow q) \leftrightarrow [p \ and \ (p \rightarrow not \ q)]$$

is a tautology. Similarly, using the second table above,

$$not(p \ and \ q) \leftrightarrow (not \ p \ or \ not \ q)$$

is a tautology. (Compare this with Theorem 8-7.)

Definition 8-4

Two propositional forms x and y with the same variables are said to be **equivalent** whenever $x \leftrightarrow y$ is a tautology.

Exercises 8-8 to 8-12

8-8. Show that each of the following is a tautology.

a. $not(p \rightarrow q) \leftrightarrow [p \ and \ (p \rightarrow not \ q)]$

b. $(p \ or \ q) \leftrightarrow (q \ or \ p)$

c. $[p \ or \ (q \ and \ r)] \leftrightarrow [(p \ or \ q) \ and \ (p \ or \ r)]$ (There are eight rows in the truth table.)

d. $not(p \ or \ q) \leftrightarrow (not \ p \ and \ not \ q)$

e. $(not \ p \ or \ q) \leftrightarrow (p \rightarrow q)$

f. $[(p \rightarrow q) \ and \ not \ q] \rightarrow not \ p$

g. $p \ or \ not \ p$

8-9. Construct a truth table for each of the following and show that it is equivalent to one of the basic propositional forms of Definition 8-2.

a. *not*(p *and not* q)

☆b. ($p \rightarrow q$) *and* ($q \rightarrow p$)

c. (p *and* q) *or* (*not* p *and not* q)

d. $p \leftrightarrow (p \rightarrow q)$

8-10. Show that each of the following is a tautology.

a. [p *and* ($p \rightarrow q$)] $\rightarrow q$

b. ($p \rightarrow q$) $\leftrightarrow$ (*not* $q \rightarrow$ *not* p)

c. [($p \rightarrow q$) *and* ($q \rightarrow r$)] $\rightarrow$ ($p \rightarrow r$)

d. [$p \rightarrow$ (q *or* r)] $\leftrightarrow$ [(p *and not* q) $\rightarrow r$]

e. [$p \rightarrow$ (q *and* r)] $\leftrightarrow$ [($p \rightarrow q$) *and* ($p \rightarrow r$)]

8-11. Can a propositional form be construed as a function (mapping) in the sense of Definition 1-10? If so, what are the domain and range?

8-12. Give examples to justify the truth values assigned in Definition 8-2 to p *and* q and p *or* q.

▲

The following discussion will center around attempts to interpret the tautologies of Exercise 8-10, and other tautologies, for the "practical" purpose of proving theorems.

Most mathematical theorems take the form $p \rightarrow q$ (if p is true, then q is true), so the most important thing to consider is a collection of ways of showing $p \rightarrow q$.

First consider the tautology ($p \rightarrow q$) $\leftrightarrow$ (*not* $q \rightarrow$ *not* p) of Exercise 8-10*b*. The fact that this is a tautology means that *not* $q \rightarrow$ *not* p is another version of $p \rightarrow q$. Thus if we can show that whenever q is false, then p is also false, we will have shown that if p is true, then q is true. Actually we are saying nothing more profound than that "If he works, he will succeed," and "If he doesn't succeed, he's not working," convey the same information. For example, we can prove that $A \subset B$ by proving either that $x \in A$ implies $x \in B$ or by proving that $x \notin B$ implies $x \notin A$. (The propositional form *not* $q \rightarrow$ *not* p is called the *contrapositive* of $p \rightarrow q$.) The reader might compare Theorem 1-1 where $E_a \neq E_b$ implies $E_a \cap E_b = \varnothing$ was proved by showing $E_a \cap E_b \neq \varnothing$ implies $E_a = E_b$. Also the first part of the proof of Theorem 5-3 is a contraposition argument.

Next consider the tautology $[(p \rightarrow q)$ and $(q \rightarrow r)] \rightarrow (p \rightarrow r)$ of Exercise 8-10*c*. This can be referred to, for obvious reasons, as the form of "proof by steps." A more elaborate version would be

$$[(p \rightarrow p_1) \textit{ and } (p_1 \rightarrow p_2) \textit{ and } \cdots \textit{ and } (p_n \rightarrow r)] \rightarrow (p \rightarrow r)$$

Almost all proofs, including all proofs which are calculations (such as those of Section 8-*A*), use this technique. It is possible to combine this technique with contraposition to get a sequence of implications beginning with *not* r and terminating with *not* p to prove $p \rightarrow r$.

A proposition such as Theorem 5-10 falls into the form of $p \rightarrow (q$ *or* $r)$ which, according to Exercise 8-10*d*, can be proved by showing that if p is true and q false, then r is true. Exercise 8-10*e* says that a way of proving a theorem of the form $p \rightarrow (q$ *and* $r)$, such as Theorems 2-4 and 2-10, is to think of it as two theorems $p \rightarrow q$ and $p \rightarrow r$. In fact, in most cases it might as well be two theorems unless both conclusions follow from a single argument or calculation.

An idea distantly related to contraposition is proof by *contradiction.* A basic form of this comes from the tautology

$$(p \rightarrow q) \leftrightarrow [(p \textit{ and not } q) \rightarrow \textit{not } t]$$

where t represents a proposition which is always true (takes on the value 1 only). That is, it is possible to show that $p \rightarrow q$ is true by taking p to be true and q to be false and showing that some proposition known to be true is contradicted.

Nothing has been said yet about how one decides what to try to prove. Actually a mathematician will occasionally examine a proof and observe that its formal structure implies a conclusion in addition to the one given. More often, mathematical propositions are formulated as a result of analyzing examples and making guesses as to what might be true—much as the experimental scientist formulates his tentative conclusions. In most research mathematics, in other words, informal guessing precedes formal proving. It is in formulating propositions and in attempting to answer questions like "Is it true?" and "If it is true, can it be proved with fewer hypotheses?" and "If it is false, what additional hypotheses—if any—will make it true?"

that the research mathematician finds much of his challenge and satisfaction.

It should be noted that mathematicians disagree among themselves as to the nature of their efforts. One extreme view is that mathematics is a game played according to rules of logic, and the conclusions reached are merely new positions in the game. Another extreme view is that mathematical results are rational statements about a rational universe and are, therefore, statements that one can make about "real" objects and apply to "real" situations in science, engineering, and life in general. The holders of both of these views accept as a basic tenet of their work the tautology

$$[p \text{ and } (p \to q)] \to q$$

of Exercise 8-10*a*. It gives both groups a reason for trying to prove $p \to q$. The difference comes from the interpretation of the word "true" in: "If p is true, and if p implies q, then q is true." That is, is a mathematical truth merely a logical consequence of a system of axioms or can it, and should it, be a friend and associate of physical truth, economic truth, and other truth?

Very few mathematicians feel either that mathematics is a pure intellectual game unrelated to the real world or that mathematics must be immediately applicable to specific scientific problems to be important. Most mathematics is produced by people who hold a view of mathematics somewhere between the extremes.

Exercises 8-13 to 8-17

8-13. Show that the propositional forms of (a) and (b) below are equivalent to $p \to q$.

a. $(p \text{ and not } q) \to \text{not } t$, where t takes only the value 1.

b. $(p \text{ and not } q) \to \text{not } p$.

c. Discuss (b) as a form of proof by contradiction.

8-14. Show that $(\text{not } p \to p) \leftrightarrow p$ is a tautology.

8-15. Discuss Exercises 8-9*b* and *c* as methods of proving $p \leftrightarrow q$.

☆*8-16.* Discuss the relationship of the propositional form $[(p \to q) \text{ and } q] \to p$ to the inductive method of experimental science. Is this a tautology?

8-17. Identify the Boolean algebra symbols $\vee$, $\wedge$, $'$, and $=$ with *or*, *and*, *not*, and $\leftrightarrow$, respectively. Identify s with a proposition t that takes only the value 1, and identify n with a proposition f that takes only the value 0. Show that symbolic logic is a Boolean algebra by showing that the logical equivalent of each of the axioms of Definition 8-1 is a tautology. (The equivalents of two of these eight axioms are in Exercises 8-8*b* and *c*.)

8-C A NOTE ON SET THEORY

We have already observed (preceding Theorem 8-5) that the collection $P(X)$ of subsets of a set X forms a Boolean algebra and (Exercise 8-17) that symbolic logic is a Boolean algebra. If we let t and f represent propositions which take only the values 1 and 0, respectively, then the correspondence between a Boolean algebra B, the set $S = \{p,q,r, \ldots\}$ of propositional variables, and $P(X)$ is given by the following table.

B	S	$P(X)$
$\vee$	*or*	$\cup$
$\wedge$	*and*	$\cap$
$'$	*not*	$\sim$
$=$	$\leftrightarrow$	$=$
n	f	$\varnothing$
s	t	X

The correspondence between S and $P(X)$ can be carried still farther. If we let p denote the proposition $x \in P$, q denote the proposition $x \in Q$, and so on, each of the tautologies involving $\leftrightarrow$ becomes a proposition about sets. For example,

$$(not\ p\ and\ not\ q) \leftrightarrow not(p\ or\ q)$$

corresponds in this way to

$$(x \notin P\ and\ x \notin Q) \leftrightarrow not(x \in P\ or\ x \in Q)$$

which corresponds to each of the statements

$$(x \in \sim P \textit{ and } x \in \sim Q) \leftrightarrow \textit{not}(x \in P \textit{ or } x \in Q)$$

$$(x \in \sim P \cap \sim Q) \leftrightarrow x \in \sim (P \cup Q)$$

$$\sim P \cap \sim Q = \sim (P \cup Q)$$

Comparing the first and the last of these, the logical form as a tautology, and the corresponding set statement, we see that a set statement is effectively proved by appealing to the corresponding tautology. As another example, the set equation

$$P \cup (Q \cap R) = (P \cup Q) \cap (P \cup R)$$

corresponds to the propositional form

$$[p \textit{ or } (q \textit{ and } r)] \leftrightarrow [(p \textit{ or } q) \textit{ and } (p \textit{ or } r)]$$

and is true because the propositional form is a tautology. A complete proof of this set equation would be obtained by using for p, q, and r the propositions $x \in P$, $x \in Q$, and $x \in R$ and invoking the tautology at a key point. Proving this one in detail:

$$\begin{aligned} x \in P \cup (Q \cap R) &\leftrightarrow x \in P \textit{ or } x \in Q \cap R \\ &\leftrightarrow x \in P \textit{ or } (x \in Q \textit{ and } x \in R) \\ &\leftrightarrow (x \in P \textit{ or } x \in Q) \textit{ and } (x \in P \textit{ or } x \in R) \\ &\leftrightarrow x \in P \cup Q \textit{ and } x \in P \cup R \\ &\leftrightarrow x \in (P \cup Q) \cap (P \cup R) \end{aligned}$$

The reader should be wondering in the midst of this discussion what became of $\rightarrow$ and $\subset$ when the correspondence was established. If he recalls the definition of $P \subset Q$ by if $x \in P$, then $x \in Q$, he will know that these are corresponding symbols. Actually, their use can be avoided whenever they occur because $p \rightarrow q$ is equivalent to (*not* p) *or* q, which means that $P \subset Q$ iff, for all x, $x \notin P$ or $x \in Q$. Thus $P \subset P$ is the set statement corresponding to the tautology $p \rightarrow p$. Tautologies involving two occurrences of $\rightarrow$ or $\leftrightarrow$ are very awkwardly interpreted in set theory, and we refrain from discussing them here—except to say that only a partial translation into set notation is possible in general.

The material of this section also raises the question of whether or not it is possible to prove Boolean algebra theorems using truth tables —which possibility we also refrain from discussing, leaving resolution of the question to the reader.

Exercises 8-18 to 8-19

8-18. For each of the following set statements write the corresponding propositional form. If the propositional form is a tautology, prove the set statement by any method. If the propositional form is not a tautology, show by a Venn diagram that the set statement can be false. [Note that $A \sim B = A \cap (\sim B)$.]

a. $A \cup B = [(A \sim B) \cup (B \sim A)] \cup (A \cap B)$
b. $A \cup (B \cap C) = A \cap (B \cup C)$
c. $(A \sim B) \cup (A \cap B) = A$
d. $(A \cup B) \cap (B \cup C) \cap (C \cup A)$
$$= (A \cap B) \cup (B \cap C) \cup (C \cap A)$$
e. $A \cup B \subset A \cup C$
f. $A \cup (A \cap B) = A$
g. $A \cap B \subset A \cup B$
h. $A \sim (B \sim C) = (A \sim B) \cup (A \cap C)$
i. $A \cap \sim B = \sim[(\sim A) \cup B]$

☆*8-19.* Define a propositional form "$p \mid q$" by $(p \mid q) \leftrightarrow$ (*not p or not q*). Show that each of the five basic propositional forms of Definition 8-2 is equivalent to a propositional form constructed using only the symbol "|" and variables p and q.

APPENDIXES

I SELECTED GROUP TABLES

Table 1. A group F of mappings (Exercise 1-63)

	a	b	c	d	e	f
a	a	b	c	d	e	f
b	b	a	e	f	c	d
c	c	d	a	b	f	e
d	d	c	f	e	a	b
e	e	f	b	a	d	c
f	f	e	d	c	b	a

Table 2. ***The group*** $\langle P(\{1,2,3\}),\Delta\rangle$
(Exercise 2-27)

Δ	$\varnothing$	1	2	3	12	13	23	123
$\varnothing$	$\varnothing$	1	2	3	12	13	23	123
1	1	$\varnothing$	12	13	2	3	123	23
2	2	12	$\varnothing$	23	1	123	3	13
3	3	13	23	$\varnothing$	123	1	2	12
12	12	2	1	123	$\varnothing$	23	13	3
13	13	3	123	1	23	$\varnothing$	12	2
23	23	123	3	2	13	12	$\varnothing$	1
123	123	23	13	12	3	2	1	$\varnothing$

Table 3. ***The group*** S ***of symmetries of the square (Exercise 2-31)***

	I	R	R^2	R^3	H	V	P	O
I	I	R	R^2	R^3	H	V	P	O
R	R	R^2	R^3	I	P	O	V	H
R^2	R^2	R^3	I	R	V	H	O	P
R^3	R^3	I	R	R^2	O	P	H	V
H	H	O	V	P	I	R^2	R^3	R
V	V	P	H	O	R^2	I	R	R^3
P	P	H	O	V	R	R^3	I	R^2
O	O	V	P	H	R^3	R	R^2	I

Table 4. The "clock arithmetic" group $\langle J_8, +_8 \rangle$

$+_8$	0	1	2	3	4	5	6	7
0	0	1	2	3	4	5	6	7
1	1	2	3	4	5	6	7	0
2	2	3	4	5	6	7	0	1
3	3	4	5	6	7	0	1	2
4	4	5	6	7	0	1	2	3
5	5	6	7	0	1	2	3	4
6	6	7	0	1	2	3	4	5
7	7	0	1	2	3	4	5	6

Table 5. The group $\langle \{1,2,4,7,8,11,13,14\}, \cdot_{15} \rangle$ (***Exercise 2-42***)

$\cdot_{15}$	1	2	4	7	8	11	13	14
1	1	2	4	7	8	11	13	14
2	2	4	8	14	1	7	11	13
4	4	8	1	13	2	14	7	11
7	7	14	13	4	11	2	1	8
8	8	1	2	11	4	13	14	7
11	11	7	14	2	13	1	8	4
13	13	11	7	1	14	8	4	2
14	14	13	11	8	7	4	2	1

II THE GENERAL ASSOCIATIVE LAW

This section presents a proof that if an operation is associative, then all products of the same elements in the same order, however associated, are equal. The operation will be denoted throughout by juxtaposition.

Definition

Let $\langle A, \quad \rangle$ be an algebraic system. The **general product** of elements of A is defined inductively as follows:

1. $\prod_{i=1}^{1} a_i = a_1$

2. $\prod_{i=1}^{n} a_i = (\prod_{i=1}^{n-1} a_i)a_n \qquad \text{if } n > 1$

Thus the general product of $a_1,a_2, \ldots ,a_n$ is the product "associated to the left." For example, $\prod_{i=1}^{2} a_i = a_1a_2$, $\prod_{i=1}^{3} a_i = (a_1a_2)a_3$, and $\prod_{i=1}^{5} a_i = \{[(a_1a_2)a_3]a_4\}a_5$. By Definition 1-16 the operation is associative in A iff $(ab)c = a(bc)$ for all $a,b,c \in A$.

Theorem

Let $\langle A, \quad \rangle$ be an algebraic system in which the operation is associative. Then every product of the elements $a_1,a_2, \ldots ,a_n$ of A, in that order, however associated, is equal to the general product $\prod_{i=1}^{n} a_i$.

PROOF. The proof, in response to the definition of general product, proceeds using the "weaker" form N' of mathematical induction discussed in Section 1-F.

The theorem is true if $n = 1$ since the general product $\prod_{i=1}^{1} a_i = a_1$ is the only one defined. Similarly, if $n = 2$, $\prod_{i=1}^{2} a_i = a_1a_2$ is the only product of a_1 and a_2, in that order, defined. [Also the theorem is true by hypothesis if $n = 3$ since $\prod_{i=1}^{3} a_i = (a_1a_2)a_3$ and $(a_1a_2)a_3 = a_1(a_2a_3)$, the only other possible association.]

Assume that every product of k elements, with $k \leq m$, is equal to the general product $\prod_{i=1}^{k} a_i$ of those elements, and let p be a product of the $m + 1$ elements $a_1,a_2, \ldots ,a_{m+1}$. Then there must be some last multiplication indicated by the association. Assume that this last multiplication is indicated between a_{r-1} and a_r so that the product has the form

$$p = (a_1 \cdots a_{r-1})(a_r \cdots a_{m+1})$$

where $(a_1 \cdots a_{r-1})$ and $(a_r \cdots a_{m+1})$ have some internal association.

Both of these products contain fewer than $m + 1$ elements so, by the induction hypothesis, $a_1 \cdots a_{r-1} = \prod_{i=1}^{r-1} a_i$ and $a_r \cdots a_{m+1} = \prod_{i=r}^{m+1} a_i$. Since $\prod_{i=r}^{m+1} a_i = \left(\prod_{i=r}^{m} a_i\right) a_{m+1}$,

$$p = \left(\prod_{i=1}^{r-1} a_i\right)\left[\left(\prod_{i=r}^{m} a_i\right) a_{m+1}\right] = \left[\left(\prod_{i=1}^{r-1} a_i\right)\left(\prod_{i=r}^{m} a_i\right)\right] a_{m+1}$$

But the product in brackets at the right contains m factors. By the induction hypothesis

$$p = \left(\prod_{i=1}^{m} a_i\right) a_{m+1}$$

which is by definition the general product $\prod_{i=1}^{m+1} a_i$. ||

III NOTES ON SELECTED EXERCISES

This is a collection of hints and complete or incomplete solutions to some of the exercises of the text—those indicated by a star. Before giving these hints and solutions, we pause to interject a few general comments on statements in proofs. These comments can be read intelligently once the definition of an operation (Definition 1-14) has been understood.

If $\circ$ is an operation in a set A, then it is a mapping and, as a mapping, it is single valued. Thus if $b = c$ (if b and c are names for the same element of A), then the ordered pair (a,b) equals the ordered pair (a,c) and there can be only one image of the pair $(a,b) = (a,c)$ under the mapping $\circ$. Thus if $b = c$, then $a \circ b = a \circ c$, and we have justified both "substitution of equals for equals" and "'$\circ$ multiplying' equals by equals." However, one must be careful, unless the operation $\circ$ is commutative, that the multiplication of $b = c$ by a take place on the same side in both members of the equation; that is, $b = c$ implies both $a \circ b = a \circ c$ and $b \circ a = c \circ a$, but does not imply $a \circ b = c \circ a$. The statement "$a \circ b = a \circ c$ implies $b = c$" (a form of cancellation) may not be true, depends on the structure of the system, and requires further justification.

It goes without saying that equality (=) is an equivalence relation (Definition 1-7). In particular, equality is transitive in the sense that $A = B$ and $B = C$ imply $A = C$. It is frequently convenient in performing calculations to adopt the format of a chain of equal things of the form $A = B = C = \cdots = F = G$. Such a chain will always be intended as a statement that the proof proceeds in that order, with each equality being justifiable, in turn, by a reason. Thus the chain is a proof that $A = G$ whose steps are $A = B$, $B = C$, . . . ,$F = G$, with the transitivity of equality implicit in the calculation.

The promised hints and solutions follow.

1-13c. To prove: $A \cap (B \cup C) = (A \cap B) \cup (A \cap C)$. Let $x \in A \cap (B \cup C)$. Then $x \in A$ and $x \in B \cup C$. If $x \in B \cup C$, then $x \in B$ or $x \in C$. Thus we have x in A and in either B or C. If $x \in B$, then $x \in A \cap B$; but if $x \in C$, then $x \in A \cap C$. Since either $x \in A \cap B$ or $x \in A \cap C$, it follows that $x \in (A \cap B) \cup (A \cap C)$. Therefore

$$A \cap (B \cup C) \subset (A \cap B) \cup (A \cap C)$$

Now let $y \in (A \cap B) \cup (A \cap C)$. Thus $y \in A \cap B$ or $y \in A \cap C$. In either case, $y \in A$. If $y \in A \cap B$, then y is also in B; if not, then $y \in C$. Thus $y \in B \cup C$. Therefore $y \in A \cap (B \cup C)$ and we have proved

$$(A \cap B) \cup (A \cap C) \subset A \cap (B \cup C)$$

1-32. Let $z \in C$. We wish to show that there exists $x \in A$ such that $x(\alpha\beta) = z$. Since β is a mapping onto C, there exists $y \in B$ such that $y\beta = z$. Since α is a mapping onto B, there exists $x \in A$ such that $x\alpha = y$. Thus, for this x,

$$x(\alpha\beta) = (x\alpha)\beta = y\beta = z$$

1-37b. A set A is finite iff $A = \varnothing$ or there exists a one-to-one mapping from A onto a set of the form $\{1,2,3, \ldots, n\}$.

1-63. As an example, $x(dc) = (xd)c$

$$= [1/(1 - x)]c = 1 - 1/(1 - x) = x/(x - 1) = xf$$

Thus $dc = f$ in F. See Table 1 of Appendix I for the completed table.

1-75. We want to prove property N, so let M be a set of nonnegative integers which contains 0 and contains $n + 1$ whenever it contains n. We shall show that $M = J_0$ by showing that $X = J_0 \sim M$ is empty. Since $0 \in M$, $0 \notin X$. If $X \neq \varnothing$, then X contains a smallest element x by well-ordering. Since $x \neq 0$, $x - 1$ is in J_0. But since x is the smallest member of X, $x - 1 \in M$. However, by hypothesis, since $x - 1 \in M$, then $(x - 1) + 1 = x$ is also in M. This contradiction of the location of x resulted from the assumption that $X \neq \varnothing$. Thus $X = \varnothing$.

1-76. *Hint:* Consider $a0 = a(0 + 0)$.

1-77. *Hint:* Consider $a[b + (-b)]$ and use Exercise 1-58.

1-84. We shall prove the first part. Let $E_c = E_a$. Thus $a \equiv_n c$ by Theorem 1-1. Since $b \equiv_n b$, $a + b \equiv_n c + b$ by Exercise 1-83, which implies that $E_{a+b} = E_{c+b}$.

2-5. *Hint:* Apply the hypothesis to the product $(a \circ b) \circ (a \circ b)$.

2-6. **1.**

$a' \circ a$	$= e$	G3
$(a' \circ a) \circ a'$	$= e \circ a'$	
$a' \circ (a \circ a')$	$= a' \circ e$	G1 and Theorem 2-1, part 2
$a \circ a'$	$= e$	Theorem 2-1, part 1

2. Similar to part 1 of Theorem 2-1.

3. $b \circ a = e = a' \circ a$ implies $b = a'$ by part 2.

2-15. *Hint:* Adapt the proof of Theorem 2-7.

2-27. See Table 2 of Appendix I for the table for $P(\{1,2,3\})$.

2-31. See Table 3 of Appendix I.

2-54. The table initially appears as follows:

	e	b	c	d
e	e	b	c	d
b	b	e		
c	c			
d	d			

The remaining entries in the second row (b row) must be d and c, respectively. The remainder of the table is determined by $cc = e$ or $cc = b$.

2-58. The conclusion of Theorem 2-10 should read

(1) $(-n)x = -(nx)$, (2) $mx + nx = (m + n)x$, and (3) $n(mx) = (nm)x$

2-64. *Hint:* If $(ab)^n = e$, compute and simplify $b(ab)^n a$.

2-68c. (1825)(37)

2-69c. (14)(253)

2-76. *Hint:* Since S_n is a group, multiplying every element of S_n on the left by (12) yields every element of S_n.

2-80a. Let $n = 2$. The elements of D_2 must be e, x, y, and xy. But completing a table for these will force $xy = yx$. (Compare Exercise 2-54.)

3-7. The mapping is $\alpha : J_n \to C_n$, where $C_n = \{e,a,a^2, \ldots, a^{n-1}\}$ is a cyclic group of order n, such that $r\alpha = a^r$. Then

$$(r +_4 s)\alpha = a^{r+s} = a^r a^s = (r\alpha)(s\alpha)$$

The mapping is clearly one-to-one.

3-13. *Hint:* The set $\{-1,1/2,2\}$ is mapped onto itself by each of the six mappings of F. Let the elements of S_3 be permutations of $\{-1,1/2,2\}$ instead of $\{1,2,3\}$.

3-15. Onto: If $\tau_a \in T$, then $a\alpha = \tau_a$. One-to-one: If $\tau_a = \tau_b$, then $x \circ a = x\tau_a = x\tau_b = x \circ b$ for every $x \in G$. By left cancellation, $a = b$.

3-26. Let the order of G be p, a prime. Let $x \neq e$, the identity of G. The cyclic subgroup $[x]$ must have order greater than 1 but cannot have as its order any of the numbers 2, 3, . . . , $p - 1$ since none of these divides p. Thus the order of $[x]$ is p, and x generates every element of G.

3-37. The definition gives closure in G/H since the product of cosets is a coset. Associativity of $\square$ is easily checked, as are the facts that $eH = H$ is an identity and that $a^{-1}H$ is the inverse of aH.

3-39b. *Hint:* Using Lagrange's theorem, generate an AB with four elements in an appropriate group of order 6.

3-45c. If K has index 2 in G, then $G/K = \{K,aK\}$ where $a \notin K$. Thus $\bar{G}$ is a two-element group (isomorphic to $\langle J_2,+_2\rangle$) and α maps elements of K to $\bar{e}$ and all elements of $G \sim K$ to the other element of $\bar{G}$.

4-3.
$$\begin{aligned} a(b - c) &= a[b + (-c)] \\ &= ab + a(-c) = ab + [-(ac)] = ab - ac \end{aligned}$$

4-12. By hypothesis, there exists an $a \neq 0$. Then $a0 = 0$ by Theorem 4-1, and $ae = a$ since e is a unity element. If $e = 0$, then $a0 = ae$, which is equivalent to $0 = a$ (a contradiction).

4-16. *Hint:* See the proof of the first part of Theorem 3-8.

4-46. Using the binomial theorem, which holds in any integral domain,

$$\begin{aligned} (a + b)^p &= a^p + \sum_{n=1}^{p-1} \frac{p!}{n!(p - n)!} a^{p-n}b^n + b^p \\ &= a^p + p\left[\sum_{n=1}^{p-1} \frac{(p - 1)!}{n!(p - n)!} a^{p-n}b^n\right] + b^p \\ &= a^p + 0 + b^p \end{aligned}$$

since D has characteristic p.

4-62b. Hint: The isomorphism is

$$a + bi + cj + dk \leftrightarrow \begin{bmatrix} a + bi & c + di \\ -c + di & a - bi \end{bmatrix}$$

4-71. Hint: Let $x = y = z = 0$.

4-73. Hint: $xd = x(00) = (x0)0$.

5-3. Hint (for part d): A may have zero-divisors.

5-10c. Hint: Look ahead to the factor theorem (Corollary 5-7a) and use the fact that $2 + \sqrt{5}$ is a root iff $2 - \sqrt{5}$ is a root.

5-24b. $q = 2x^2 + 2$, $r = 2$.

5-32. Proof for $n > 0$ and $d > 0$: Let $M = \{n - dq \mid q \in J_0$ and $n - dq > 0\} \subset J_+$. Thus M has a smallest element

$$n - dq_0 = r_0$$

since $M \neq \varnothing$ $(n - d0 = n \in M)$. If $r_0 > d$, then

$$0 < r_0 - d = (n - dq_0) - d = n - (q_0 + 1)d$$

Thus $r_0 - d \in M$, contradicting the choice of r_0. If $r_0 = d$, then $n = dq_0 + d = d(q_0 + 1) + 0$, which satisfies the theorem. Otherwise, $r_0 < d$. The cases of $n \leq 0$ and/or $d < 0$ can be handled using this case. The uniqueness can be proved in ways similar to Theorem 5-5.

6-3. Hint: Try Table 3 of Appendix I. Keep Exercise 3-31 in mind.

6-18. Define α by $(g,h)\alpha = gh$. Thus α maps $G \times H$ onto (by condition 5) GH. The mapping α is readily shown to be operation-preserving. Also $(g,h)\alpha = e$ iff $gh = e$ iff $g = h^{-1}$ iff

$$g,h \in G \cap H$$

iff $g = h = e$. Thus the kernel of α is $K = \{(e,e)\}$, so α is an isomorphism.

6-29. Hint: This follows quickly from the "$g^{-1}Hg \subset H$" characterization of normality.

6-32. Since a subgroup K is nonempty, $g^{-1}Kg \neq \varnothing$. If $g^{-1}ag$ and $g^{-1}bg$ are in $g^{-1}Kg$, compute $(g^{-1}ag)(g^{-1}bg)^{-1}$ and use Theorem 2-6.

6-37. *Hint:* It is sufficient to show that if $x = ab \in AB$, then every conjugate of x is in AB.

7-11. *Hint:* See Theorem 4-1.

7-18b. If $a_1y + x_1$ and $a_2y + x_2$ are in K and if $b,c \in F$, then

$$b(a_1y + x_1) + c(a_2y + x_2) = (ba_1 + ca_2)y + (bx_1 + cx_2)$$

is in K since $bx_1 + cx_2 \in H$.

7-27. If $\alpha,\beta \in \mathfrak{L}(G,L)$, define $\alpha + \beta$ by $x(\alpha + \beta) = x\alpha + x\beta$ and define $a\alpha$ by $x(a\alpha) = a(x\alpha)$ for all $x \in G$. With these definitions $\mathfrak{L}(G,L)$ can be shown to be a vector space.

7-36. The proof that $k(A + B) \subset k(A) + k(B)$ follows from Theorem 7-6. *Hint:* To show that $k(A) + k(B) \subset k(A + B)$, show first that $A + k(B) \subset k(A + B)$, then replace A by $k(A)$.

7-46b. *Hint:* Use the equation

$$\frac{1}{a + b}(x + y) = \frac{a}{a + b}\left(\frac{1}{a}x\right) + \frac{b}{a + b}\left(\frac{1}{b}y\right)$$

to show $(x + y)\beta_A \leq a + b$, and take infima on the right.

8-9b. The last column reads 1,0,0,1, which is the same as the truth table for $p \leftrightarrow q$.

8-16. *Hint:* If you perform an experiment several times and always get the result q, does q result *because of* the initial conditions p of the experiment?

8-19. Two of the answers are: $(p \mid q) \mid (p \mid q)$ and $p \mid (q \mid q)$.

INDEX OF SYMBOLS

SUBJECT INDEX